W E W

A Field Guide To:

Southern

California

Third Edition
Robert P. Sharp
California Institute of Technology

KENDALL/HUNT PUBLISHING COMPANY
4050 Westmark Drive P.O. Box 1840 Dubuque, Iowa 52004-1840

Copyright © 1976, 1979, 1994 by Kendall/Hunt Publishing Company

ISBN 0–8403–8437–8

Library of Congress Catalog Card Number: 93–61226

Printed in the United States of America
10   9   8   7   6   5   4   3   2   1

Contents

Foreword

Dr. Sharp's book is one of a series of books in the Kendall/Hunt *Geology Field Guide Series.* The objective of this series is to provide an authoritatively written layperson's guide to important geologic features in each region treated in the series. The books stress observations in the field. Each guide provides an overview of the geologic provinces of the region to which it pertains and outlines a series of self-guiding field trips that will allow users to make their own firsthand observations on features that typify the provinces.

The series is directed toward diverse groups of users. The series should find use in formal classes in geology, both at the college and university levels, and in high schools, in which field trips form an essential part of introductory or advanced courses. Furthermore, books in the series should be useful to professional geologists and other scientists who desire an introduction to the geologic features of particular regions. Finally, the series should find use among individuals who are not necessarily trained in science, but who do have an active interest in natural history and who enjoy travel.

Authors of books in this series all have intimate acquaintance with their respective regions and extensive teaching experience which has stressed field trip observations. Consequently, each book in this series represents a distillation of teaching experiences that have involved many students and numerous field trips.

John W. Harbaugh
Consulting Editor

Preface

Most things are interesting, even old bleached cow bones, provided you know something about them. Indeed, much of the pleasure and enjoyment of life come from knowledge, familiarity, and understanding of things, be they related to sports, music, drama, science, art, birds and bees, or flowers and trees. Among scientists, geologists are reputed to have as much fun as anybody because of their understanding and appreciation of the natural environment. An objective of this effort is to share that fun with you.

Consisting of four chapters, this book is designed for people without any formal acquaintance with geology but with an inherent interest in nature and the out-of-doors. The first chapter aims to provide some background in basic geological matters; it can well be skipped by those with previous knowledge of the subject. Chapter 2 furnishes descriptions of geological features and relationships in nine natural provinces within southern California. The third and fourth chapters, constituting well more than half the book, are guides to geological features visible along two principal travel routes, one from Los Angeles Basin to Death Valley and the other from Los Angeles Basin to Mammoth up the east side of the Sierra Nevada. Guides for supplementary spurs to the Las Vegas area, Palm Springs, and San Joaquin Valley are provided in Chapter 4. Emphasis is on geology that can be seen while you travel by car along highways, often at a fair rate of speed.

Despite current interest in the moon and other planets, most of us are going to spend all of our life on the surface of planet Earth, and we will enjoy our stay more if we know something about aspects of the natural environment. A speaking acquaintance with the surrounding rocks, geological structures, and landforms can be a source of real satisfaction and enjoyment. This book aims to excite your interest in these matters and to impart some information about them at the same time.

Use of technical jargon is minimal, but avoidance of all geological terms is impossible. A foreign language cannot be read without at least some vocabulary; so a glossary is provided for necessary terms (Appendix B). A book of this size cannot possibly cover all the geology of southern California, so great selectivity had to be exercised. This is not an "all about" book, but it is hoped you will find it a fun book. Most of you already know more geology than you realize.

I am deeply indebted to Dorothy L. Coy and Carolyn Porter for skillful and patient typing and retyping of manuscript material; to Enid H. Bell for critical reading and editorial suggestions; to Janice I. Mayne and Ruth Z. Talovich for excellent drafting services; to John S. Shelton, Roland van Huene, Pierre St. Amand, Robert C. Frampton, and Malcolm M. Clark for help in procuring illustrations; to Helen Z. Knudsen for many custom photographs; to Kathleen G. Nelson for manuscript proofreading; and to Joe and Clem Frindt for layperson reactions. Colleagues in the Division of Geological and Planetary Sciences of the California Institute of Technology have provided many bits of geological information. Bruce T. Sharp assisted in the construction of field trip guides; and Frank and Irene Goddard, Mack and Enid Bell, and Don and Eileen Burnett provided helpful evaluations of them.

Background

Most of us take the ability to read written material for granted, forgetting how we struggled to learn as small children. Understanding geology through observation of natural features is simply another form of reading. It, too, requires a little knowledge and some practice, but the effort is well worthwhile. Reading of words is essential to safe and comfortable existence in the modern world ("Stop," "Danger," "Women," "Men"), and the reading of nature's geological record makes that existence more enjoyable.

The principal sources of geological information are rocks and landforms. Both have been around for a long time and have fascinating stories to tell, if you just learn to ask questions in the right way and to listen for the answers carefully. Mostly, it's a matter of being interested and of gradually accumulating experience. If you like natural landscapes, you can soon learn to understand at least something of the history of past events that they tell.

The characteristics of landforms are determined principally by three factors: the nature and structure of underlying materials, the surface processes at work, and the stage of development attained in the interaction between processes and materials. A fourth factor important in landscape development—particularly in southern California—is natural deformation of the earth's surface. In a sense, this deformation is the starting point of most landscape evolution.

MATERIALS

Minerals are the basic units of earth materials. They are solid natural substances of relatively fixed chemical composition and characteristic physical properties such as hardness, luster, and color. The ice in your refrigerator is a mineral, as is the salt you sprinkle on french fries. Look at table salt with a magnifying glass to see its beautiful little cubic crystals. Many people enjoy collecting minerals, especially gems. Some durable minerals, such as zircon, are prized by geologists because they retain a record of time and events extending over billions of years.

1

Photo 1–1.
Dark dikes intruded into massive white sandstone, near Point Mugu, Malibu coast.

A rock is an aggregate of minerals, either in grains or crystals, although some—such as rock salt or marble—consist of a single mineral. Rocks are the principal constituent of the earth's crust, and they have been used by humans from time immemorial. One of the first tools our far-distant ancestors used was a rock: for settling an argument with a neighbor, subduing a spouse, or killing a wild animal. Rocks are the greatest recorders of history the world has ever known.

You are already familiar with some rocks: the marble of a soda-fountain counter, the granite facing of a bank building, the slate of a black-board, the sandstone or schist of your fireplace or patio. Rocks come in three principal classes, and you should know their names: *igneous, sedimentary,* and *metamorphic.* It may help your memory to know that students sometimes speak in jest of "ingenious," "sedentary," and "metaphoric" rocks. Don't expect to recognize all rocks instantly. Even experienced geologists are not above using the cryptic notation *frdk,* for "funny rock don't know."

Igneous rocks were once molten and were able to intrude other rocks (Photo 1–1). Upon cooling they crystallized into aggregates of minerals. Sedimentary rocks are secondary, composed of rock or mineral fragments derived from the disintegration and decomposition of preexisting rocks. They can also consist of mineral matter deposited from chemical solutions or of organic substances like plant remains (coal). Metamorphic rocks are also secondary. They are formed by modifications produced in igneous and sedimentary rocks by pressure or heat, or both, sometimes aided by chemical solutions. As a result of such modification, new physical characteristics are developed that are wholly unlike those of the original rock. For example, squeezing the living daylights out of

Photo 1–2.
Strong foliation (banding) in a gneissic rock.

shale, a sedimentary rock, compacts it and gives it a special facility for splitting into smooth thin sheets (*rock cleavage*) that makes it the metamorphic rock, *slate*. If the temperature is high enough and significant recrystallization occurs, new minerals are formed and a *schist* is produced. Because igneous and metamorphic rocks are usually composed of aggregates of mineral crystals, they are often spoken of as *crystalline rocks.*

Igneous rocks, other than volcanics, tend to be mostly massive and homogenous. Sedimentary rocks are characteristically layered or bedded. Metamorphic rocks may be either, but in addition, many display an irregular lamination known as *foliation* (Photo 1–2), most typically seen in *gneiss* (pronounced "nice"). Kent Clark, lyricist laureate of the Caltech campus,

once composed a small ditty to gneiss, part of which goes something like this:

> Limestone and coral
> Can never be moral
> They're not gneiss

Here is a simple rock table with a few common rock names in each class. See the glossary, Appendix B, for more complete descriptions.

Igneous	Sedimentary	Metamorphic
Granite	Conglomerate	Gneiss
Granodiorite	Sandstone	Quartzite
Diorite	Siltstone	Schist
Gabbro	Shale	Slate
Lavas { Basalt Rhyolite	Limestone	Marble

Photo 1–3.
A syncline and an anticline in soft Tertiary beds west of the parking lot at Calico, field trip Segment B.

GEOLOGICAL STRUCTURES

To most people the solid earth seems dead, because they observe too small a part of it for too short a time. Actually, it is very much alive. It takes energy to keep things living, and the earth gets its energy from disintegration of internal radioactive substances. This produces heat, some of which is dissipated by conduction to the surface and some of which is converted into mechanical energy that bends and fractures rocks. To a geologist, these bends and fractures are *structures;* specifically, they are folds and faults. An upfold is an *anticline;* a downfold is a *syncline,* or a "sinkline" in jocular terms (Photo 1–3). Fractures are *joints* (Photo 1–4), but when significant slippage occurs (or has occurred) along a fracture, it becomes a *fault*. Such movements cause earthquakes, and southern California is richly endowed with both faults and earthquakes. As the late Romeo Martel, dean of earthquake structural engineers, used to say, "California with all your faults we love you still, only you don't stay still long enough."

Photo 1–4.
Large granitic boulder split along a joint plane formed in the parent bedrock. (Photo by Wakefield Dort, Jr.)

Photo 1–5.
The Cristianitos fault, pale massive sandstone against darker fractured shale exposed at the foot of sea cliff in San Onofre State Beach Park, San Diego County. This is a hands-on structure; you can go right up and pat it.

Geological faults of southern California are of two types. Those with principally sidewise or lateral movement, and those with up-down movement. We accordingly speak of lateral faults (more properly, lateral-slip faults) and up-down faults.

If you stood on one side of a fault when lateral slip was occurring, the other side would appear to be moving past, either to the right or left. Think about this for a moment and you will realize it doesn't make any difference which side you stand on, the sense of displacement appears the same, either right or left. Hence come the terms *right-lateral* and *left-lateral* fault. Both sides of a fault may actually have moved in the same direction but one more than the other. In speaking of up-down faults (Photo 1–5), observers say that one side has gone up or down relative to the other be-

cause both may actually have gone up or down, again, one more than the other. Actually, fault movements are commonly somewhat oblique, not strictly lateral or up-down. Finally, the fracture surfaces (planes) of some faults are so gently inclined that they are closer to horizontal than vertical. These are called *thrust faults,* on which one side is shoved over the other, or *low-angle up-down faults,* on which one side moves off the other.

California's greatest fault is the San Andreas (see Photos 2–4 and 2–16). It slashes 650 miles across the state from the Mexican border to Cape Mendocino, 225 miles north-northwest of San Francisco (Figure 1–1). This is a classic example of an active, northwest trending, right-lateral

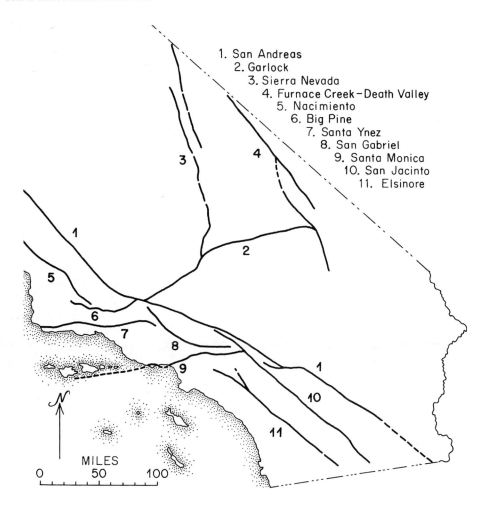

1. San Andreas
2. Garlock
3. Sierra Nevada
4. Furnace Creek–Death Valley
5. Nacimiento
6. Big Pine
7. Santa Ynez
8. San Gabriel
9. Santa Monica
10. San Jacinto
11. Elsinore

Figure 1–1.
Generalized fault map of southern California.

fault. Others of this type in southern California are the San Jacinto, Elsinore, and probably the Newport-Inglewood (see Figure 2–7).

Geological relationships suggest that the total right-lateral displacement on the San Andreas has been hundreds of miles. Displacements on the San Andreas system during the last 100 years average about two inches per year. Because Los Angeles is west of the fault and San Francisco is mostly east of it, these two cities get a little closer together each time the fault slips. The San Andreas or its associated branches are responsible for three of the major interior gateways to the Los Angeles region, namely San Gorgonio, Cajon, and Tejon passes. Many of you have crossed the San Andreas dozens of times.

A different type is represented by east-west faults with predominantly up-down movement, such as the Foothill (see Figure 2–7) at the south base of the San Gabriel Mountains. The north side of this fault has moved relatively up creating the steep south face of the San Gabriels. This face or *scarp* has a vertical relief of 7,000 feet in the vicinity of Cucamonga and Etiwanda peaks. The Banning fault along the south base of the San Bernardino Mountains is a member of this same family.

There are thousands of smaller faults in southern California that do not correspond in trend or sense of displacement with these major types. Furthermore, the Garlock (see Photo H–1), a huge fault extending from Tejon Pass near Lebec to the south end of Death Valley, bears east-northeast and has a large *left* lateral displacement.

Southern California has so many fractures that a fault map of the region looks like an Italian glass mosaic (see Figure 2–7). Various pieces within this mosaic have moved up, down, or sidewise, helping to create the varied topography that characterizes the area. The magnetic orientation of minerals within some of the rocks composing the individual pieces, largely lavas, suggest they have also rotated by as much as 90° in the last few million years.

Nearly all sections of sedimentary rocks, more than a few thousand feet thick, in southern California are tilted (Photo 1–6) or folded. They are also faulted, because faults have no respect for rock types. Only rarely in this domain do you see near-horizontal bedding, and when it does occur, it is almost invariably a mark of geologically young deposits.

Many of our topographic prominences such as Signal Hill, Dominguez Hills, Santa Monica Mountains, and Wheeler Ridge (see Photo 2–17), among others, are primarily the result of folding. We recognize upfolds (anticlines or oppositely inclined limbs) more readily than downfolds (synclines or limbs inclined toward each other) because the former make ridges or ranges, and the latter make valleys that become filled with rock, sand, and dirt (*alluvium*). San

Photo 1–6.
Exposed bedding planes in tilted, thinly layered Cozy Dell shale, Highway 33, upper Matilija Creek, north of Ojai.

Fernando Valley is basically a large synclinal structure lying north of the anticlinal Santa Monica Mountains. The valley extends five times farther down into the earth than the Santa Monica Mountains extend up. It is easy to think too much of our mountains and to forget that our valleys are also geologically significant (see Figure 2–3).

Massive igneous and metamorphic rocks are subject to the same forces that cause sedimentary beds to fold. Usually they respond by fracturing and faulting, but in some instances they warp. This means that the rocks bend or tilt gently on a broad scale. Warping is an important geological phenomenon, but the results are often hard to recognize. If it weren't for the layering in sedimentary rocks, it might not be obvious that they had been folded.

Southern California is different from much of the rest of the nation in that the folding and faulting have occurred so recently that much of the topographic relief is due directly to such movements. High-standing areas have been folded, warped, or faulted upward, and low-standing areas the reverse. This is not true in other parts of the country, such as the Appalachian Mountains of Pennsylvania, where the deformation occurred long ago. There the present relief was caused primarily by differential erosion of hard and soft masses of rock.

The recency and variety of geological structures are what endow California with its unusual terrain, including the highest and lowest spots of our contiguous states. California rocks are not all that different from those in some other parts of the country, and surface processes acting on these rocks are about the same as elsewhere. We simply live in a geologically dynamic area, and it shows. In subtle, psychological ways the youthful geology may be partly responsible for the fact that southern Californians are something of a race apart.

PLATE TECTONICS

Plate tectonics is one of the most fruitful concepts yet developed within the earth science. It states that the surficial crust of planet Earth consists of eight major plates, and several smaller plates, that move with respect to each other and from place to place over the planetary surface. Many of you will relate this to the age-old concept of continental drift. Plate tectonics confirms that continents do indeed drift, and ocean basins do, too. This concept has provided a logical explanation of how and why the movements happen.

Earth, like an apple, has a thin skin. Geologists call it the *crust,* beneath which is a layer 1,800 miles thick, the *mantle,* that corresponds to the fleshy part of the apple. At the center of both apple and Earth is a core, the Earth's having a radius of 2,200 miles. Under ocean basins the Earth's crust is much thinner—mostly about 3 miles—than it is under continents, where it can be as much as 30 miles thick (20 miles is average). Plates are still thicker, about 50 miles, be-

cause they include part of the underlying mantle. Beneath plates is a weak mantle layer, several hundred miles thick, that deforms easily in a plastic mode. This layer is what enables plates to move at speeds ranging from a fraction up to 4 inches per year; 2 inches is about average.

Major plates can be huge, consisting of an entire continent and a large part of an adjacent ocean basin. The North American plate, for example, extends from the middle of the Atlantic Ocean to the Pacific Coast. The Pacific plate is also huge, but it consists almost entirely of oceanic domain.

Plates are separated by three types of boundaries. One is the large, broad, submerged, oceanic ridges where plates are growing as molten material wells up from the mantle to form new oceanic crust, which pushes older crust outward on both sides. These are called *rises* or *spreading ridges.* Another boundary is a long narrow zone on the sea floor, commonly bordering a continent, where slabs of plate material plunge down into the mantle. These are termed *subduction zones,* and their course is usually marked by a deep sea *trench,* some up to 35,000 feet deep. They form where two plates move together head-on and develop best where one plate consists of lighter continental rocks and the other of heavier oceanic material. The oceanic side is the one that plunges, and in so doing it usually creates a lot of igneous activity, both intrusive and extrusive (volcanic) along the margin of the over-riding continental plate. Astrophysicists and geophysicists tell us that our planet is currently neither growing nor shrinking perceptibly. Thus formation of new crust at spreading ridges must be balanced, on the average, by consumption of old crust at subduction zones.

The third type of plate boundary is the one dominating southern California. It is a huge complex fault zone along which two plates slip past each other. In California that zone is the famed San Andreas fault, which slices generally north-northwesterly through the southern two-thirds of the state. These boundaries are called *transforms,* and they develop where two plates move obliquely so their movements can be resolved into a lateral displacement. This resolution, as you can imagine, creates considerable stress and deformation within the margins of the plates. Southern California straddles the zone between the southwest-moving North American plate and the northwest-moving Pacific plate. Hence the region is beset by earthquakes and has many faults and folds. For simplicity's sake we speak of plate boundaries as being ridges, trenches, or transforms.

This account of the plate tectonics concept is but a bare-bones digest, because many complications exist. We need to consider one, known as a *triple junction,* where three plates join. Triple junctions involve any combination of plate boundaries, but one is commonly a transform. The three plates can be moving, in different directions at different speeds. Consequently, a triple junction is a point of stress and, furthermore, it moves on its own. In the last 25 million years, a triple junction moved northwesterly along and through the coastal part of southern California, spreading tectonic havoc along the way, much like a geological cyclone of very slow speed. All this makes southern California an interesting—even exciting, but not necessarily the safest—place to live.

Geological opinion on the cause of plate movements is still in flux. We know that more heat is generated within the mantle by radioactive disintegration than can be disposed of by conduction. Therefore, it is postulated that the mantle has large, slow-moving convection cells. Convection currents passing beneath plates could conceivably cause some movement. This is termed *mantle drag.* Plates are also pushed outward from spreading ridges as new crust forms, and the great height and breadth of such ridges would help them along. This has been called *ridge push.* Finally, downward-plunging slabs at subduction zones are thought to pull plates along. Crustal rocks are so brittle, however, that they can't stand much pull. Perhaps subduction simply reduces the resistance to movement. It is known that some fast-moving plates are subject to both ridge push and trench pull. Whatever the mechanisms, it is obvious that mantle processes have a lot to do with what happens to the crust on which we live. Plate tectonics has helped us appreciate that.

Photo 1–7.
Cavernous weathering in coarse sandstone near base of sea cliff on Highway 1 west of Topanga Canyon.

SURFACE PROCESSES

So much for materials and structures. What are the processes working on them to carve out the landscape? They are numerous but do not all act in concert or with the same degree of effectiveness in different environments. Basically, we are dealing with the interaction of the solid Earth with the atmosphere, hydrosphere, and biosphere—including humans. Humans modify the earth's surface with all of their chicken scratches. People just haven't been here long enough to carve out a Grand Canyon. However, for their size, humans have proven to be efficient dirt movers.

Weathering

If you leave your car out in the open long enough, it will start to rust and disintegrate. That's *weathering,* the reaction of solid substances with the atmosphere and biosphere. Rock surfaces have been exposed to the elements for hundreds of thousands of years; consequently, essentially all of them display some degree of weathering (Photo 1–7). Most geologists didn't know what truly fresh unaltered rock looked like until they saw chunks of moon rock. Weathering is a pervasive, night-and-day,

Photo 1–8.
Scar of an earthflow
in central Puente
Hills, viewed
northeastward.
(Photo by John S.
Shelton, 720.)

omnipresent process. It works slowly, but over long periods of time it is very effective. Without it you and I wouldn't be here, because weathering produces one of our necessary natural resources: soil. Given time and proper conditions, weathering will make a fine soil out of a solid mass of granite. The chemical and physical changes produced in rocks by weathering play a major role in landscape development by making rocks susceptible to erosion.

Mass Movements

Once rocks have been softened and disintegrated by weathering, the debris starts to move. Initial downslope movements are caused by gravity unaided by other agents of transport. The material simply creeps, slides, rolls, or flows downslope. This is mass movement.

Some types of mass movement are so slow that they are imperceptible in short-term observations. If you seated the statue of Venus de Milo on a soil-mantled hillside, you would probably find over a period of 10 years that the good lady would tilt a bit and move downslope a foot or two. This behavior results from the insidious process of *creep* that prevails to some degree on practically every debris-mantled slope.

Earthflows (Photo 1–8) and *mudflows* (Photo 1–9) are forms of mass movement that occur episodically and usually at an easily observable rate. *Landslides* move with even greater velocity and are fully capable of engulfing anyone in their path. Some, like the rock-fall slide (Photo 1–10) that took place near Blackhawk Canyon on the north face of San Bernardino Mountains many thousands of years ago, probably attain a velocity well in excess of 100 miles an hour. The lobate tongue of broken rock in the photograph traveled 4.3 miles out over the gently sloping desert floor at the east end of Lucerne Valley before coming to rest.

Photo 1–9.
Small house at Wrightwood buried to the eaves by a muddy debris flow in 1941.

Photo 1–10.
View south over breccia lobe of Blackhawk rockfall slide on north side of San Bernardino mountains east of Lucerne Valley. Toe of lobe is 4.3 miles from mountain base. (Photo by John S. Shelton, 2459.)

Photo 1–11.
Arroyo Seco, near Devils Gate dam, Pasadena, in flood stage. The Rose Bowl sits on the normally dry floor of the Arroyo a mile or so downstream. So far so good, thanks to a flood-control channel, but a larger flood could be embarrassing.

Running Water

The most effective agent of erosion on earth is running water. Any high-standing land mass is subjected to its attack, except in unusual places, such as Antarctica, where the water is frozen. Running streams carve valleys, canyons, gorges, arroyos, barrancas, and gullies, all integral parts of our landscape. It's easy to underestimate the effectiveness of a stream until it goes into flood (Photo 1–11). Then skeptics quickly become believers, because streams are terrifyingly powerful in flood stage. Even on southern California deserts, water is the principal agent of erosion and transport because it is not impeded much by vegetation; hence, flash floods are common.

Running water is not wholly a destructive agent. Through *deposition* it also acts constructively. Some of the most heavily populated parts of the southern California landscape—alluvial fans and alluvial plains—consist of debris deposited by streams.

Wind

Wind is localized in its geological effectiveness, but where it works, it's effect is great. Wind erodes, transports, and deposits (Photo 1–12). Wind-blown sand is a highly effective agent of abrasion (Photos S-7, 8), as anyone emerging from a Coachella Valley sandstorm with a frosted windshield can testify. In addition to blasting windshields, wind-blown sand undercuts power line poles and etches and polishes stones resting on the ground, converting them to *ventifacts* (literally, "windmade"). They are not windmade, of course, only wind-shaped.

Photo 1–12.
Wind ripples in dune sand. The larger crescentic forms are in coarser sand.

Most agents transport material downhill. Wind is one of the few agents that can and does carry material uphill. In deserts, winds carry sand for tens of miles before piling it up into huge dunes hundreds of feet high. The shapes and forms associated with sand dunes are some of the loveliest in nature, especially when seen in low evening or morning lighting. That's the witching time in dunes.

Glaciers

The San Bernardino Mountains harbored seven small valley glaciers clustered in the San Gorgonio Peak area during the last ice age, about 15,000 years ago. However, the Sierra Nevada Mountains display the erosional and depositional products of mountain glaciers at their best. The broad U-shaped valleys, glacial steps, waterfalls, hanging valleys, polished rock surfaces, bedrock-basin lakes, and huge piles of rocky debris, known as *moraines,* left by the glacier are as well displayed there as anywhere in the nation.

Shoreline Processes

The principal source of energy at the earth's surface is radiation from the sun. About 70 percent of the earth is covered by oceans that intercept the major share of this solar radiation. Oceans are therefore great pools of energy, some of which they expend along their contact with land, the seashore.

Owing to the constant attack of waves and currents, the seashore is geologically dynamic. There's a lot of erosion, transportation, and deposition going on, and the scene is ever changing. Sea cliffs retreat at rates measured in many feet per year (Photo 1–13), and beaches change aspect with every storm, and regularly with the season. Winter storms move sand off California beaches, leaving piles of boulders. In summer, waves and currents bring sand back, burying the boulders and restoring the beaches. The balance of sand supply and sand movement along beaches is a delicate thing, as humans have learned time and time again to their sorrow when they have interfered with shoreline processes. Our current damming of flood waters in streams on land is reducing the amount of sand supplied to the ocean. This practice is currently having a strong adverse effect on our beaches. A more complete statement concerning shoreline processes is made in the Kendall/Hunt companion book to this one entitled *Coastal Southern California*.

Photo 1–13.
Sea cliff and wave-cut platform at Laguna Beach, Orange County.

GEOLOGICAL TIME SCALE

Don't be overawed by the carefree way geologists speak of millions and billions of years. Humans have been on this good earth for such a few ticks of the geological clock, which started running 4.6 billion years ago, that it's hard for us to put geological events into proper perspective. If the entire span of geological time were compressed into 1 year, and we imagined the present to be exactly midnight of December 31, one of my associates, George Clark, calculates that the Declaration of Independence was signed 1.4 seconds ago, Columbus discovered America 3.3 seconds ago, Christ was born about 14 seconds ago, and 100 seconds ago ice covered much of Europe and North America and our Stone-Age ancestors huddled in caves. Human beings appeared on Earth at about 7:15 this evening, and living creatures first crawled out of the water back in the last week of November.

Knowing this, you can appreciate that our personal, first-hand experiences with earth activities, processes, and history cannot be anything but incomplete. That's why geologists are so interested in talking with rocks 2 or 3 billion years old. Those rocks are wise in the ways of the world.

For our purposes, relative age relationships are also important. Geologists have built up a relative time chart filled with curious names; a simplified version is supplied in Appendix A. Don't try to memorize this chart; rather, look at it from time to time as you come across geological-time names in these pages. Gradually you will learn that something classed as Cenozoic is relatively young and that Precambrian rocks or events are relatively ancient. The geological time scale is your chart for navigating the vast seas of geological history. Don't hesitate to use it.

Natural Provinces of Southern California

People have mental limitations, and nature is infinitely complex. To deal with this situation people invent classifications. Nature does not classify trees, flowers, and rocks—humans do, so that we can deal with them in a reasonable fashion. In order to facilitate geological understanding, southern California is divided into nine provinces (Figure 2–1). A *natural province* is a region with characteristics that distinguish it from other regions. It can be defined on the basis of vegetation, climate, topography, or the color of jack rabbits. We use geology.

The California Division of Mines and Geology sells a dandy small-scale geological map of California. It shows natural provinces, major faults, and a generalized distribution of rock types. Appendix C provides directions for obtaining a copy.

TRANSVERSE RANGES

New arrivals, and even a few "aborigines" (residents of more than 10 years), in the greater Los Angeles-Santa Barbara area, easily become confused on compass directions, and with good reason. "Up the coast" is commonly taken to mean north, but it's not north in this province. In traveling a straight course from San Bernardino to Santa Barbara, one ends up 137 miles west and only 27 miles north. To go north from Los Angeles you travel almost directly into the interior toward Mojave, crossing east-west mountains in the process.

Character and Dimensions

This east-west landscape defines a geological province, the Transverse Ranges, so named because they are crosswise to the usual northwesterly fabric of California. This characteristic is established by faults and folds that control the trend and shape of mountains, valleys, and the coastline. Even the San Andreas fault trends more east to west within the Transverse Ranges than normal. This kink is

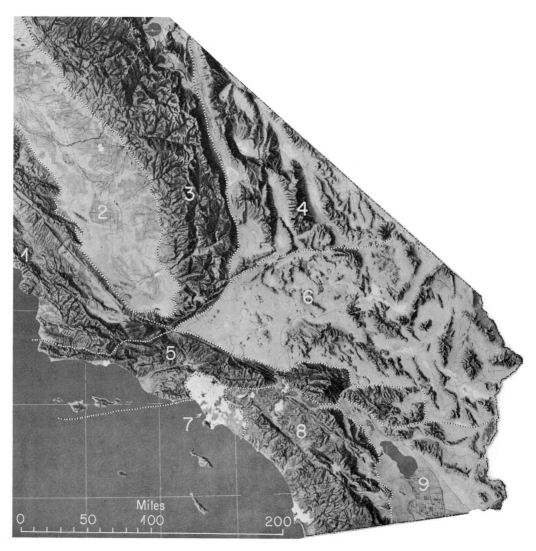

Figure 2–1.
Natural provinces of southern California. Base Map © Jeppeson and Co., Denver Colorado. All rights reserved.

colloquially known as the "Big Bend." The cause of this unusual orientation has long puzzled geologists. Simple north-south compression could have caused both the east-west folds and faults and the "Big Bend" in the San Andreas, but the source of such north-south compression remains a subject of speculation. It is probably related in someway to movements of the gigantic Pacific and North American plates toward and past each other, but their respective northwest and southwest movements are not likely to produce north-south compression.

Modern geophysical research is showing that Earth's upper mantle is much more heterogeneous than earlier thought. Geologists have come to realize that conditions and events in the upper mantle strongly influence behaviors

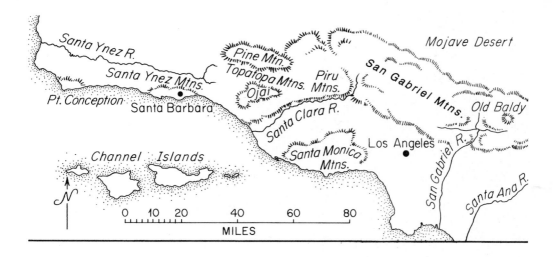

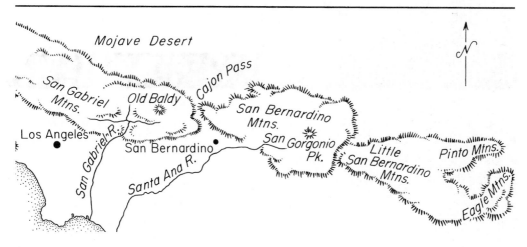

Figure 2–2.
Transverse Ranges province, west half above, east half, with overlap, below.

in the overlying crust. Possibly, the unusual structural trends within the Transverse Ranges reflect an anomaly in the underlying mantle that resolves the forces created by the impingement of the Pacific and North American plates into a local north-south compression. Just how that could happen is not clear.

Topographically the Transverse Ranges are a rugged chunk of country featuring high, rough mountain masses and long, narrow, intervening valleys. Many of the higher peaks of southern California, outside the southern Sierra Nevada, are in this province: San Gorgonio at 11,499 feet and San Antonio (Old Baldy) at 10,063 feet, are two.

The principal ranges and valleys of this province are identified in Figure 2–2. The Channel Islands are included because they are but a partly submerged westward extension of the Santa Monica Mountains. This province averages 30 miles

Photo 2–1.
Typical Transverse Ranges country, interior Santa Ynez Mountains. Santa Ynez River and Santa Barbara's Gibraltar dam and reservoir are in the foreground. (Spence air photo E-8000, courtesy of Department of Geography, University of California, Los Angeles.)

wide and is nearly 300 miles long, extending from Point Arguello, 55 miles west of Santa Barbara, eastward at least to Eagle Mountains in the Colorado Desert.

Rocks

It is convenient to think of this province as divided into eastern and western parts by the Golden State Freeway (Interstate 5) between Los Angeles and the San Joaquin Valley. In the west, sedimentary rocks predominate; in the east, older igneous and metamorphic rocks are the rule. Volcanic rocks are locally represented in both parts. The Santa Ynez (Photo 2–1), Topatopa (Photo 2–2), Piru, and Pine (Photo 2–3) mountain ranges of the west (see Figure 2–2) display sequences of sedimentary rocks, aggregating tens of thousands of feet in thickness. The San Gabriel, San Bernardino, and other ranges farther east are a mixture of

older crystalline igneous and metamorphic rocks, including some of the most ancient in southern California, at roughly 1.7 billion years. The intervening valleys are underlaid by thick accumulations of sedimentary beds or by crystalline and sedimentary rocks buried under unconsolidated deposits of sand, gravel, and dirt, called *alluvium*.

Structure

Essentially everyone agrees that this province has been subjected to strong compression, which has produced tight folds in the sedimentary rocks as well as high- to low-angle faults. These folds and most of the faults bear east-west establishing the characteristic linear topographic pattern of mountain ridges and intervening valleys. These topographic features in turn control the pattern of highways and the location and shape of urbanized areas.

Photo 2–2.
Skyline made by Eocene strata of Topa Topa Bluffs in Topa Topa Mountains of Transverse Ranges.

Photo 2–3.
Skyline ridge is Pine Mountain in western Transverse Ranges of northern Ventura County. Outcropping rocks primarily Eocene.

Folds are most abundant to the west where sedimentary sections are thickest. Some are huge: Witness the large syncline underlying Santa Clara Valley in Ventura County (Figure 2–3). High-angle faults are also numerous in folded areas, for example the San Cayetano and Oak Ridge faults bounding the Santa Clara Valley syncline.

Massive crystalline igneous and metamorphic rocks yield to stress more by fracturing than by folding, so faults are dominant in the largely crystalline-rock eastern Transverse Ranges. A number of faults are inclined northward at relatively high angles, and the north side of most has moved relatively upward, creating what geologists call *reverse faults*. The Foothill fault along the south base of San Gabriel Mountains is a good example. Movement on it caused the disastrous 1971 San Fernando earthquake and probably the gentler 1991 Sierra Madre shock. The San Gabriel fault running along the backbone of these mountains is a member of a different fault family. It shows evidence of many miles of right lateral displacement and is possibly an inactive, ancestral member of the San

Andreas fault system. After helping align the east and west branches of the San Gabriel River, it flares off northwesterly toward Castaic and Frazier Mountain.

Recently, some investigators have postulated that the Transverse Ranges consist of a stack of rock slabs shoved one on top of another by faulting, with a large, near-horizontal fault about 10 miles deep at the base. The Whittier Narrows earthquakes of 1987 might have been generated by such a fault. This is an intriguing speculation. What we now need is more earthquakes to provide further data—or, perhaps, safer and possibly cheaper—deep drill holes.

The principal dissonant note in the east-west fabric of the Transverse Ranges is the San Andreas fault (Photo 2–4), and its close relative, the San Jacinto, which slash through the province in a slightly more northwesterly direction, particularly in the Cajon Pass-Lytle Creek area. These are large active faults with many miles of right-lateral displacement. The San Gabriel and San Bernardino mountains would make a continuous range if they were

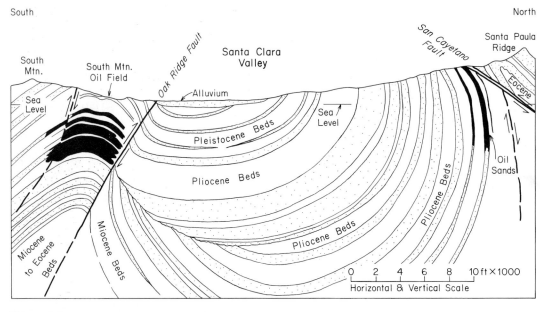

Figure 2–3.
Geological cross section of Santa Clara Valley.

not sliced apart by these gigantic ruptures. Flying out of Los Angeles east-bound via Cajon Pass, you can often see off the left wing the great gash of the San Andreas cutting discordantly across the terrain. It looks as though the Jolly Green Giant had hacked a trench across the land with a huge machete.

Resources

Much oil (Photo 2–5) has been produced from the western Transverse Ranges where the many folds and some associated faults provide traps for its accumulation. Petroleum geologists long ago discovered that oil, which is lighter than water, tends to migrate upward into anticlines, where it accumulates beneath the curving cap of some impervious bed. If the impervious cap is cracked by faults and fractures, oil may seep through to the surface. This has been happening for thousands of years in the Santa Barbara Channel. A leaky anticline and poor drilling practice were basically the reason for the disastrous oil slicks associated with operations in the area in 1969.

If oil is king in the western Transverse Ranges, iron is queen of the east, where Eagle Mountains supplied much of the ore for the now-defunct Kaiser's Fontana mills. Limestone is quarried for cement, some building stones are produced, sand and gravel are exploited, and a little placer gold has been recovered from the East Fork of San Gabriel River. A water-laid accumulation of rock debris—usually sand and gravel containing fragments of precious mineral—is a placer deposit.

An unusual nonmetallic resource of the province is diatomite, largely from the Lompoc area of western Santa Barbara County. Diatoms are tiny single-celled organisms, halfway between a plant and an animal. They require sunlight, water, silica, and nutrients such as phosphorous, carbon dioxide, and a little iron.

Photo 2–4.
View west-northwest over Palmdale reservoir and along San Andreas rift up Anaverde and Leona valleys. Antelope Valley to right with Portal Ridge between. (Spence air photo E-14483, 01/26/54, courtesy Department of Geography, University of California, Los Angeles.)

Photo 2–5.
Oil seep over road cut on Highway 150 along Sisar Creek, about two miles west of Sulphur Springs.

They love cold, calm water (salt or fresh) and are most abundant in polar oceans. Like plants, they employ sunlight to photosynthesize CO_2 into carbohydrates with oxygen as a by-product, using a couple of dyes in place of chlorophyll. About 60 percent of our atmospheric oxygen comes from the ocean, most of it produced by diatoms. We might not be here were it not for these tiny critters. A diatom reproduces itself once in about every 24 hours.

Some 5 to 10 million years (m.y.) ago the ocean over the area around Lompoc had conditions highly favorable to diatom propagation. As a result, more than 3,000 feet of sediment accumulated on the sea floor consisting largely of diatom skeletons. Volcanic emanations and volcanic ash may have provided the silica needed.

Diatom skeletons are small but extremely beautiful and complex geometrically, so each skeleton has an incredibly large surface area for its size. Consequently, accumulations of skeletons have a low bulk density, about 0.1 or 0.2, and there are about 20 million of them in a cubic inch of diatomite. It has been said that a pound of diatomite has a surface area equivalent to 10 football fields. The Lompoc region has the world's greatest accumulation of diatomite, currently mined from huge open pits. Lompoc has enough diatomite to supply the world's needs for several hundred years.

Diatomite is reported to have more than 100 uses. Being composed of silica, it is physically stable and chemically nonreactive, so it makes a good filler in many substances and is used in

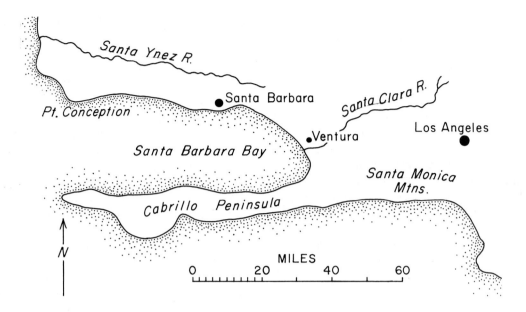

Figure 2–4.
Sketch of postulated Cabrillo Peninsula.

ceramics and insulation. Most of all, it makes superb filters for a wide variety of fluids, from motor oil to beer.

Special Features

Cabrillo Peninsula. A few million years ago the structural uplift, represented by Santa Monica Mountains, may have extended far westward as a long narrow peninsula converting the present Santa Barbara Channel into a bay (Figure 2–4). This peninsula, named in honor of the great Portuguese explorer, Juan Rodriguez Cabrillo, encompassed the present Channel Islands of Anacapa, Santa Cruz, Santa Rosa, and San Miguel, where Cabrillo may be buried. Structurally, these islands are anticlinal with numerous associated faults, one

of which on their south side appears to be the far-western extension of the Santa Monica fault (see Figure 1–1).

Cabrillo Peninsula was inhabited by mainland plants, such as Torrey pines, and animals, including elephants that were then reasonably abundant in California. Eventually, perhaps a million years ago or less, the Cabrillo Peninsula subsided, leaving only its higher points projecting as the present islands. The elephants were isolated, and in the course of time evolved into a species of pygmies standing only 6 to 8 feet high, compared to a normal elephant height of 10 to 13 feet. This change is attributed to insular isolation, but just how or why isolation on an island breeds smaller animals is not wholly clear. Maybe the herd had to huddle in caves, and only the smaller ones could get in or only small elephants got enough food.

Some investigators hypothesize that there was no Cabrillo Peninsula and that the elephants swam to the islands. Elephants can swim, but what motivated them to swim, perhaps as much as 20 miles, to the islands when the mainland provided more food and territory? Further, why didn't pygmy elephants swim back to the mainland? Their fossil remains are found only on the islands. The answers may be that although the channel between mainland and islands might have been only 4 miles wide during a lowered glacial-period sea level, once the sea level rose to something near its present level, elephants, were reluctant to try a 20-mile swim and became isolated. Arguments favoring a peninsula as an avenue for migration of trees as well as elephants seem a little preferable to swimming elephants.

Sespe Formation. To geologists, a *formation* is a body of rock of consistent characteristics formed within a discrete span of time under specific conditions. One distinguishing feature of the Sespe Formation is its red color. It is not all red, but those parts that are, like red-headed girls and boys, attract attention. The Sespe is also a land-laid formation, that is, it was deposited as a series of gravel, sand, and mud layers on land rather than in the sea. This distinguishes it from thick sequences of sedimentary beds above, below, and alongside that are of marine origin. The Sespe represents about 20 million years of time, embracing the interval from late Eocene to early Miocene (see Appendix A).

The Sespe is not restricted to the Transverse Ranges, but that location is its principal abode. Remnants are preserved in many places between Point Conception and the Santa Ana Mountains. They are seen in lower Topanga Canyon, along the south face of the Santa Monica Mountains, on the Channel Islands, in Simi Valley, the Las Posas, in South Mountain opposite Santa Paula, and on the upper and lower parts of Sespe Creek, whence comes the name. As you travel up or down the Santa Clara Valley, look north to the mouth of the Sespe Canyon as you cross the Sespe Creek bridges just west of Fillmore. If the light is right, you will see dark-red sandstones east of the canyon mouth. Those are Sespe outcrops, and large reddish boulders in the stream bed are samples of the rock.

The Sespe Formation was deposited in a large land-locked basin bounded by relatively high mountains. The valleys of this basin must have been well watered, because they were inhabited by large rhinoceros-like beasts and other heavy vegetational browsers. Picture a lushly vegetated basin extending 150 miles east-west across southern California and imagine what a scene it made when populated with lemur monkeys, primitive dogs, hyena-like carnivores, camels, horses, primitive pig-like cud chewers (oreodonts), deer-like animals, tapirs that looked like an early form of small stocky horse with a pendant snout, and a variety of large rhino-like animals tromping around the premises. There were even a few primitive saber-toothed cats and lots of small rodents and rabbits. Titanotheres (titanic mammals), sort of a cross between a horse and a rhinoceros, also stalked the scene. Such was this land of ours in Sespe time; it looked more like a part of present-day Africa.

Moon-like rocks in the San Gabriels. Exposed along Mill Creek, on the Angeles Forest Highway, is a large body of light-colored (almost white), coarsely crystalline, igneous rock, called *anorthosite,* composed predominantly of large, light-colored feldspar crystals and segregated masses of dark minerals. The Apollo missions to the moon brought back rock samples from the lunar highlands that proved to be anorthosite. See field-trip Segment I for more details.

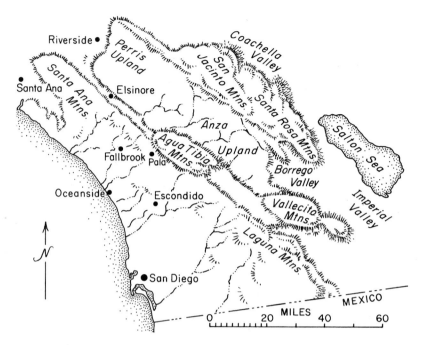

Figure 2–5.
Peninsular Ranges province.

PENINSULAR RANGES

The Peninsular Ranges in the southwest corner of southern California constitute a land of gems, telescopes, and *batholiths*. Don't recoil from that word: *bathos* = "deep," and *lithos* = "rock"; it's that simple. A batholith is an igneous intrusive body of large size that cooled at great depth so slowly that the molten material had time to crystallize into a coarse-grained rock. The southern California batholith finds its home in this province, one of the world's large optical telescopes is on Palomar Mountain in the Agua Tibia Range, and the Pala area was once a great gem-collecting area. The Peninsular Ranges actually belong more to Mexico than to the United States, because they extend along the entire 800 miles of peninsular Baja California, whence comes the name. Americans can claim only a small northern fraction of the province.

Character and Dimensions

This province has a distinct (but not overpowering) northwest grain expressed by its higher mountains and longer valleys (Figure 2–5). It also includes a lot of hilly country without a strong linear pattern, such as the area around Fallbrook. There's a hill here, a vale there, and the roads and ranches are comfortably dispersed about the landscape. This local lack of linear geometrical pattern reflects the homogeneity of the underlying batholithic rocks.

The Peninsular Ranges merge northward into the Los Angeles Basin, and the northwest grain eventually terminates against the east-west Transverse Ranges. The Peninsular Ranges are bounded on the east by the Salton Trough.

Westward, the province does not end at the Pacific shore, as one might assume, but continues far out under the ocean as a broad submerged continental borderland. The islands of Catalina, San Clemente, Santa Barbara, and San Nicholas are simply the highest parts of largely submerged peninsular-type ranges.

The U.S. portion of the province extends 130 miles north from the Mexican border. The maximum land-bound width is about 65 miles, but including the submerged continental borderland, the width is 225 miles. Some of the larger ranges and major valleys within the land-bound sector are identified in Figure 2–5. Hot springs keep things warm here and there, Warner's being the best known.

In gross aspect, the province is a large block uplifted abruptly along the eastern edge and tilted westward. The highest point, San Jacinto Peak (10,801 feet), towers more than 10,000 feet above Palm Springs and Coachella Valley. This scarp is as high as the east face of the Sierra Nevada and surpasses in vertical relief the east face of the Grand Tetons west of Jackson Hole, Wyoming.

The mountains and ranges making up this province rise steeply and abruptly above adjoining valleys, but much of the upland country has a gentle relief. The area around Palomar Observatory, the west side of the Santa Rosa Mountains along the Palms-to-Pines Highway, and the west flank of the Laguna Mountains farther south display large areas of gentle upland. The flat terrain around Perris and March Field is similar but occurs at a lower level. Faulting and homogeneous igneous rock play major roles along with erosion in producing this steep-sided, gentle-topped topography.

Much of the coastal margin has one or more wide flat benches on which the highways and coastal settlements are situated. These are *marine terraces*; the flat treads represent uplifted sea-floor platforms, and the steep slopes between are old sea cliffs. They are described in greater detail in Kendall/Hunt's *Coastal Southern California Guidebook.*

Rocks

If you dote on relatively coarse-grained, homogeneous igneous rocks, and if you like the type of soil and landscape they yield, this is your province. Rocks of the southern California batholith are dominant. Although formed at depth, they have subsequently been exposed at the surface by deep erosion. However, because these rocks had to be intruded into something older, it is not surprising to find sheaths of metamorphic rock around the margins of intrusive bodies, and within them pendant metamorphic remnants. The oldest of these metamorphics are schists, quartzites, and coarsely crystalline marbles.

A somewhat younger group of prebatholithic rocks appears originally to have been shales and volcanics. The Santiago volcanics, a thick accumulation of largely *clastic* (broken up) volcanic debris with interlayers of fine grained slates and sandstone, are well exposed in Santa Ana Mountains and in Black Mountain east of San Diego. They are of varied composition and moderately metamorphosed. The interlayered sediments contain sparse middle to late Jurassic fossils, and radiometric dates on volcanic constituents give ages as young as middle Cretaceous. They are intruded by the batholithic igneous rocks.

The batholithic rocks consist of many separate intrusive units ranging in composition from *gabbro* (dark) to *granite* (light). They are accompanied by a host of *dikes,* narrow sheetlike igneous bodies intruded into cracks. One dike rock, called *pegmatite,* is particularly important. A pegmatite is often of granitic composition, and it can be very coarse-grained with individual crystals measured in inches, sometimes feet. Pegmatites seem to be formed by the last vapors and fluids given off by a cooling and crystallizing igneous body. Rocks of the southern California batholith are mostly 70 to 120 m.y. old.

Younger rocks found in the province are largely sedimentary, partly marine and partly terrestrial, and range in age from late Cretaceous (80 m.y.) to Pleistocene (1.8 m.y. or less). Marine rocks are exposed mostly in the Santa Ana Mountains and in the belt of hills along the coast from Corona Del Mar to San Diego.

The bluffs at Torrey Pines State Preserve expose an intriguing sequence of Eocene rocks demonstrating how slow submergence of this area in the ocean about 55 m.y. ago caused clean, wave-washed sands of an offshore bar, the Torrey sandstone, to come to rest on top of fossiliferous lagoonal deposits, the Delmar Formation. The story is in the sea cliffs.

The Poway area east of San Diego is the type locality for one of California's geological mysteries. Here the 40 to 45 m.y. old Poway Conglomerate contains distinctive, smooth and well rounded, volcanic stones for which no source has yet been positively identified. These stones have been reworked into younger deposits all over the San Diego area.

Terrestrial deposits were laid down in inland basins; the Borrego Badlands expose this type of accumulation. Volcanics of about mid-Miocene age (15 m.y.) appear in some places, including Catalina and San Clemente islands, the latter being largely volcanic. Still younger lavas are preserved southeast of Lake Elsinore.

Structures

The northwest grain of the Peninsular Ranges is caused primarily by faulting. Folding is locally significant mostly in the Santa Ana Mountains and coastal hills, and parts of the province have probably been broadly warped. Major faults are the San Jacinto and Elsinore (see Figure 1–1), and both are active. These are complex structural zones accompanied by a number of parallel, individually named satellitic faults, particularly toward the southeast. A similar pattern of faults controls the topography of the offshore continental borderland.

The San Jacinto and Elsinore faults, like the San Andreas, have experienced much right-lateral displacement. Many of the other faults show considerable vertical movement, although the topographic relief along them is attributed as much to differential erosion of rocks on opposite sides of the fault as to direct displacement. In terms of seismic energy released within historical times in Southern California, the San Jacinto has been more active than the San Andreas.

Resources

Attempts have been made to mine gold, lead, zinc, copper, nickel, and tungsten in this province, but only the production of gold, largely from quartz veins in the Julian area, has been significant. Some gravels in that region contain small accumulations of placer gold.

The province is more noted for its nonmetallic products. Cement comes from marble bodies in the Jurupa Mountains near Colton and Riverside. Gypsum is mined in the Fish Creek Mountains of Imperial Valley, and building and decorative stones are quarried in the southern California batholith. Roofing granules come from several places, and sand and gravel from many.

The Paleocene Silverado Formation of the Santa Ana Mountains is unusual for its content of clays that are of good enough quality to be used in roofing tile and pipe, for its content of low-grade coal, and for medium-grade glass sand. All three are rare items in southern California, and they reflect an unusually stable, low relief terrain and a moist environment in this area about 65 m.y. ago. The clay is a product of long chemical weathering, the glass sand is a residual product of deep rock weathering, and the coal indicates a swampy setting. All three bespeak a warm, moist climate, not at all like present-day southern California.

Special Features

Gems. Since 1872 the Pala district, near the center of this province, has been one of the noted gem-mineral collecting areas of the nation. Something approaching a million dollars of gemstones have been commercially produced, mostly between 1900 and 1922, and large quantities of rare minerals and gems have been obtained for private collections. The region is particularly noted for its lithium-bearing minerals, one of which—spodumene, a lithium aluminum silicate—forms gems with exquisitely delicate pink and lavender tints. Tourmaline, the other common gemstone, is a chemically complex compound that is attractive because of its hardness, high luster, and brilliant green and pink coloration, like a ripe watermelon.

These minerals occur in pegmatite dikes of which over 400 have been mapped in the Pala district. They are irregular bodies, commonly lenticular, but occasionally with bulges 100 feet across. Pegmatites tend to be extremely coarse grained, and they contain a wide variety of unusual and highly volatile chemical elements, which is the reason for their exotic mineral composition. Mineral collectors still occasionally find interesting "pickings" in the Pala area.

Hot springs. The Peninsular Ranges have a surprising number of hot mineral springs, considering the province is not a young volcanic type. Elsinor has been noted for its hot sulphur-water baths, and Murietta and Warner hot springs were long favored spas. These spots and a handful of small local hot springs are all known or thought to be on faults, of which the Peninsular Ranges has an abundance.

LOS ANGELES BASIN

Los Angeles Basin means different things to different people. To officials of the Air Pollution Control District, it is that region all too often submerged in smog. To professional geologists, it signifies a former basin on the sea floor in which a great thickness of mud and sand accumulated. For purposes of this book, it is that part of our land extending south from the foot of San Gabriel Mountains to the sea and southeast from Santa Monica Mountains to Santa Ana Mountains and San Joaquin Hills (Figure 2–6).

Many geologists include Los Angeles Basin within the Peninsular Ranges because it has the same northwest structural grain. This book handles it separately and accords it a more detailed consideration because it is the abode of so many people and because none of the geological trip guides of this book deal with the Los Angeles Basin.

Character and Dimensions

The Los Angeles Basin is divided into a northern one-third and a southern two-thirds by the Puente-Repetto hills (see Figure 2–6). South of these hills is a lowland sloping gently toward the sea—a coastal plain. North is San Gabriel Valley and the western part of San Bernardino Valley, alluvium-filled basins encircled by hills and mountains.

Figure 2–6.
Los Angeles Basin; A-Anaheim, CM-Corona Del Mar, HB-Huntington Beach, LA-Los Angeles, LB-Long Beach, N-Norwalk, P- Pasadena, Po-Pomona, RB-Redondo Beach, SG-South Gate, SM-Santa Monica, W-Whittier. Modified from R. F. Yerkes et al., 1965, USGS Prof. Paper 420-A.

The coastal plain rises from the sea to elevations of a few hundred feet with a few scattered hills and mesas projecting 100 to 200 feet above its level; Baldwin, Dominguez, and Signal hills are examples. Palos Verdes Hills at the southwest edge interrupt the otherwise smooth transition of the coastal plain to the sea. They are like an offshore island captured by seaward growth of the land. The Los Angeles Basin is enclosed on its landward sides by mountains rising abruptly thousands of feet. The basin's alluvial filling irregularly penetrates these bounding hills and mountains like a sea encroaching upon a mountainous coastline.

Three of southern California's larger streams—the Los Angeles, San Gabriel, and Santa Ana rivers—traverse the basin. They are in large part responsible for the alluvium that now mantles its surface, and they also cut right across hills that lie across their path.

The long axis of the basin extends northwesterly fully 50 miles from the San Joaquin Hills near Laguna to the Santa Monica Mountains. The span from Palos Verdes Hills to San Gabriel Mountains approaches 35 miles.

Rocks

The surface of Los Angeles Basin is largely covered by stream-laid sand, gravel, and silt that effectively hide the underlying bedrock. However, the basin is as full of holes as high-grade Swiss cheese, these holes being the tens of thousands of wells drilled in search of oil and water. From them, and from much subsurface geophysical exploration by oil companies, a lot is known about the thickness and nature of rocks beneath the surface. However, to see these rocks in outcrops one must repair to hilly areas within the basin or to mountainous areas bordering it. The rocks of Los Angeles Basin comprise three principal groups: (a) old crystalline basement rocks, (b) prebasin sedimentary rocks, and (c) materials filling the basin, called *basin sediments* or *basin fill*.

The basement rocks are of two types, designated for convenience the eastern and western complexes. The eastern complex consists of igneous and metamorphic rocks of the types seen in the Peninsular and Transverse ranges. The western complex consists primarily of an unusual metamorphic rock, the Catalina schist. It is exposed on Catalina Island and in a small area high on the northeast flank of Palos Verdes Hill. The Catalina schist is unusual in that some of its minerals contain abnormally high levels of sodium that give them a peculiar bluish color. Exposed in sea cliffs around Laguna is a coarse bouldery deposit containing large angular fragments of this schist. The location, age, and structure of this deposit suggests that roughly 15 to 20 m.y. ago a high landmass lay just off the present coastline shedding coarse rocky debris into the Laguna area.

Younger than the basement rocks (but still older than the basin fill) are sedimentary deposits ranging from late Cretaceous (70 m.y.) to lower Miocene (20 m.y.). They are layers of conglomerate, sandstone, and shale aggregating a total thickness of nearly 17,000 feet. Included is our old friend, the Sespe Formation. The Los Angeles Basin did not exist as a separate feature when these deposits were laid down.

Subsidence of the Los Angeles depositional basin started about 15 m.y. ago in mid-Miocene time, and was accompanied by the extrusion of volcanic material. The Topanga Formation, the deposit first laid down in the basin, locally contains as much as 3,000 feet of volcanic rocks and associated near-surface intrusives within its total thickness of 10,000 feet. It's fun to reflect that not so long ago our backyard was dotted with active volcanos spouting fumes, ash, fragmented ejecta, and red-hot lava over the surroundings. These volcanic rocks can be seen today around the edge of the basin, in Palos Verdes Hills, Griffith Park, or Puddingstone State Park in the Covina Hills.

A little uplift and erosion occurred before renewed subsidence led to additional deposition of thick sections of Miocene, Pliocene, and Pleistocene sedimentary beds, largely of marine origin. In places, the Pliocene sandstone, siltstone, shale, and conglomerate alone have an aggregate thickness of 14,000 feet. Some of these deposits are exposed in the Repetto Hills (see Figure 2–6) that give their name to the Pliocene Repetto Formation.

The earliest Pleistocene deposits are marine, and some contain abundant fossil sea shells, particularly in Palos Verdes Hills. However, as the basin filled, the surface rose above the sea and terrestrial deposition followed the receding shoreline outward. One of the most unusual of these Pleistocene terrestrial deposits is the tarpit accumulation of Rancho La Brea (commonly called the La Brea tar pits). Here bones, representing the amazing animal life of southern California 12,000 to more than 40,000 years ago, are preserved in tar. By all means, visit the exhibits of Rancho La Brea fossils in the Los Angeles County Museum in Exposition Park and the displays and restorations in the George C. Page Museum at the fossil pits in Hancock Park off Wilshire Boulevard toward Sixth Street between Ogden Drive and Curson Avenue. The parking area is entered from Curson. There is little point in describing sabertoothed tigers, giant wolves, ground sloths, elephants, buffalo, lions, and the birds of Rancho La Brea when you can see their restored remains for yourself.

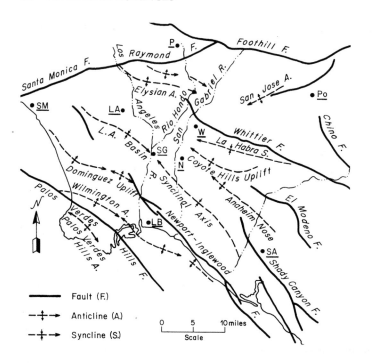

Figure 2–7.
Geological structures in the Los
Angeles Basin; A-Anticline,
F-Fault, LA-Los Angeles,
LB-Long Beach, N-Norwalk,
P-Pasadena, Po-Pomona,
S-Syncline, SA-Santa Ana,
SG- South Gate, SM-Santa
Monica, W-Whittier. Modified
from R. F. Yerkes et al., 1965,
USGS Prof. Paper 420-A.

Structure

Los Angeles Basin is a huge downfold (*syncline*) with a central line (*axis*) extending northwest from Santa Ana to Beverly Hills (Figure 2–7). The next time some world traveler regales you with stories of towering Himalayan peaks just say, "Let me tell you about the basement relief of the Los Angeles Basin." The lowest part of the basement surface lies between South Gate and Downey at 31,000 feet below sea level. The top of Mt. Wilson, only 20 miles north and composed of similar basement rock, approaches 6,000 feet above sea level. Total basement relief is thus 37,000 feet. If all the sedimentary fill were cleaned out of the basin, the topographic relief would exceed that of the Himalayas at their highest point, Mt. Everest, by nearly 7,000 feet. It would be a varied topography to boot. As you can see from Figure 2–7, the Los Angeles Basin is replete with folds and faults. Each of

these structures is represented by a major feature of relief, thousands of feet high, on the underlying basement floor.

These structures bear northwesterly, which is why many geologists include Los Angeles Basin within the Peninsular Ranges. Indeed, some of the structures actually continue from the Peninsular Ranges into the basin, but note how they are all cut off sharply on the northwest, at the junction with the Transverse Ranges, by the Santa Monica-Raymond fault. Faults determine the boundaries of Los Angeles Basin on two other sides: the Palos Verdes Hills fault on the southwest and the Foothill fault on the north. Faults also define units or blocks within the basin, the central block, for example, being bounded on the northeast by the Whittier fault and on the southwest by the Newport-Inglewood fault. Actually, major faults cutting across Los Angeles Basin divide it into four, not just two, blocks. To some degree these blocks have behaved independently, and some or all may have rotated significantly in the last few million years.

The Newport-Inglewood fault zone is a complex structure with a northwesterly trace marked by a succession of mesas and hills extending from Newport Bay to Beverly Hills, of which Signal, Dominguez, and Baldwin hills are the more prominent. The zone produces a vertical displacement in the underlying basement of nearly 4,000 feet and separates the western and eastern basement complexes, suggesting that this may be an old feature of truly major magnitude. In the overlying sedimentary rocks, the fault zone consists of a sequence of short parallel overlapping linear segments, and the displacement decreases upward so that the youngest beds are offset only 200 to 300 feet. This evidence indicates that movement on the fault has occurred repeatedly and in small increments. Old rocks record the cumulative sum of these displacements, but young rocks show only the latest movements. Other structures in the Los Angeles Basin—folds as well as faults—suggest a similar history, that is, initiation at about the time of original basin subsidence 15 m.y. ago and subsequent intermittent activity up to the present. Although possibly old, the Newport-Inglewood fault zone is anything but dead, as attested by the Capistrano earthquake of 1812 and the disastrous Long Beach shock of 1933. Geological relationships suggest considerable right-lateral displacement along the Newport-Inglewood fault, so it may be a member of the San Andreas family.

Some structures extend from the Peninsular Ranges into the Los Angeles Basin, Shady Canyon, and El Modeno faults at the southeast corner (see Figure 2–7), for example. Whittier fault also appears to be essentially a continuation of the Elsinore fault, a major structure of the Peninsular Ranges. The Whittier fault extends along the south base of the Puente Hills and dies out west of Whittier narrows in the Repetto Hills. Like the Elsinore, the Whittier fault exhibits significant right-lateral movement, but it also displays much vertical displacement. The actual movement has probably been oblique, that is, partly up and partly lateral, and it has not been inconsequential, totaling at least 15,000 feet.

San Gabriel Valley lies between Puente-Repetto hills and San Gabriel Mountains. The latter have been formed by large and recent uplift along the east-west Foothill fault. Splitting off to the southwest from the Foothill fault in the Arcadia-Monrovia area is the Raymond fault. The Huntington Hotel and the Huntington Library are perched atop a young scarp marking the trace of this fault through the Pasadena-San Marino area. Horse racing fans may enjoy knowing that the hillside stretch of the turf course at Santa Anita comes obliquely down the face of the Raymond scarp. This may well be the only section of racetrack in North America traversing a fault scarp. The fault runs westward to join the Santa Monica fault that extends along the south side of Santa Monica Mountains, past Malibu and Corral Canyon, and far westward along the south side of the Channel Islands (see Figure 1–1), there called the Anacapa fault.

Hilly areas within the Los Angeles Basin are mostly anticlines, trending principally northwesterly and plunging in that direction. Take this book, fold it into an anticline, and then tilt it so the crest is no longer horizontal. Now you have a plunging anticline. The arrow at the end of a line representing a fold in Figure 2–7 indicates the direction of plunge. Some folds are doubly plunging, that is, they plunge northwest in one place and southeast in another, Palos Verdes Hills anticline being an example. The Los Angeles Basin synclinal axis also plunges from both directions toward its lowest spot under the junction of Rio Hondo with the Los Angeles River near South Gate (see Figure 2–7).

A geologically interesting aspect of most folds in the Los Angeles Basin is their youth. They are so young that the uplifted areas have the configuration of the structure itself. Although dissected by gullies and canyons, a doubly plunging anticline still looks in profile like a doubly plunging anticline. This is not true of folds in many other deeply eroded parts of the world. Earthquakes and the measurement of geodetic markers suggest that deformation is still going on in Los Angeles Basin.

Some of you may be aware that the area around Anaheim has sunk a foot or two in the last decade and that the area around Wilmington has sunk nearly 20 feet in the last 30 years. The old Edison Company power plant at Wilmington is now below sea level. This sinkage is thought to be primarily caused by the withdrawal of fluids—oil, water, and gas—from beneath the ground. It is of artificial origin rather than a result of natural deformation.

Resources

The Los Angeles Basin, with some 46 separate oil fields, is a prolific petroleum province. Total production since about 1880 exceeds 5 billion barrels, nearly half the total for all of California. Most of the oil is pumped from Lower Pliocene and Upper Miocene rocks, and most of the fields are on anticlines.

In the nineteenth century California was the United States' major gold producer, but by the midtwentieth century it had faded to a minor contributor. Production has picked up again, owing largely to new techniques implemented in the 1980s. By 1987 the state was the nation's number two producer, behind South Dakota. During some of the down years, gravel pits near Azusa were the major source of California's gold. It was obtained as a byprod-uct of sand and gravel washing, the original source being in gold-quartz veins up the East Fork of San Gabriel River.

Special Features

Palos Verdes Hills. Palos Verdes Hills rise like an island above the alluvial sea of the Los Angeles Basin. Indeed, the hills were an island in the ocean less than a million years ago, and their bedrock basement is closely allied to that of Catalina. Structurally, the Palos Verdes Hills are a doubly plunging anticline bounded by a fault on the northeast side (see Figure 2–7). A small area of Catalina schist is exposed in the core of this anticline high on the northeast flank along the headwaters of George F. Canyon, between Crest Road and Palos Verdes Drive. A dozen smaller anticlines and synclines are exposed on the flanks of the main fold; so you will see beds inclined in just about all directions in different road cuts.

The most widely exposed rock unit is the Monterey Formation, a thinly bedded, brown-to-white shale that covers fully 90 percent of the surface. Locally it is rich in diatoms, the skeletons of microscopic single-celled plants that float around in water. A quart of sea water can contain a million diatoms. Diatom skeletons are made of silica, and where richly concentrated they compose deposits of diatomite that is used in filters and adsorbers. Diatomite was formerly quarried in the Rolling Hills district of Palos Verdes Hills.

Overlying the Miocene rocks, mostly on lower slopes around the edges, are thin deposits of younger Pliocene and Pleistocene beds. Some of these younger accumulations are richly fossiliferous, furnishing a variety of sea shells in great abundance and in a good state of preservation. The old Second Street locality in San Pedro has long been noted for its Pleistocene marine fossils.

Photo 2–6.
Uplifted, wave-cut, marine terraces at west end of Palos Verdes Hills, viewed southeastward. (Photo by R. C. Frampton and John S. Shelton, 4–9064B, 1953.)

Palos Verdes Hills are justly famous for many features, but one of their more infamous features is the landslide at Portuguese Bend. The following constitutes a recipe for disaster: Take a series of soft, thinly bedded, incompetent shale layers, incline them southward toward the sea, let the sea be constantly undercutting and steepening the slope, and add some lubrication in the form of altered clay-rich volcanic ash and water. The result—almost inevitably a landslide.

The Portuguese Bend slide had probably moved many times before urbanization of the area, but in 1956 it slipped again, destroying almost totally about 100 houses and seriously damaging at least 50 more at an estimated loss of $10 million. The courts have allowed a claim of more than $5 million against the County of Los Angeles, so all county tax payers have a stake in the problem.

The pity is that this area had obviously undergone sliding in the past, and renewed movement would have been a reasonable expectation. It is testimony to the inadequacy of our zoning practices that the area was not early designated as a park for riding trails, picnic grounds, and open-space usage. The slide moves slowly, at most a few inches per day, so recreational uses could be compatible with this gradual movement.

A striking feature of Palos Verdes Hills is the marine terraces—13 in all—rising as a succession of stair-like steps from sea level to an elevation of 1,300 feet (Photo 2–6). These are particularly prominent at the west end of the hills in the Lunada Bay area. The next to lowest terrace, about 140 feet above sea level, is the widest and best preserved. Many developments such as parks, golf courses, houses, streets and highways like Palos Verde Drive are on terrace surfaces. Higher terraces are largely obscured by burial beneath the sloping surfaces of later alluvial deposits.

These terraces demonstrate that geological deformation can occur in separate spurts. Each terrace represents a stable period of little or no deformation when the sea was cutting horizontally into the land (Photo 2–7). Such stable periods were terminated by uplifts that raised the wave-cut platform and the sea cliff at its inner

Photo 2–7.
Slide on face of Pacific Palisades blocking Pacific Coast Highway west of Santa Monica. The Palisades were formed as the ocean cut into the land creating a sea cliff. (Photo by John S. Shelton, 987.)

edge, converting them to the stair-like step we call a terrace. Thus, Palos Verdes Hills attained the present height of 1,480 feet in at least 13 separate episodes of uplift, probably more. Remember this when you look at the 8,000-foot scarp along the east face of the Sierra Nevada. This scarp wasn't created all at once; rather, it represents a summation of countless uplifts, 5 feet now, 3 feet then, and 10 feet some other time.

Antecedent streams. Many of you have driven north or south on San Gabriel River Freeway (I-605) through the Whittier Gap or Narrows, perhaps countless times. If you ever stopped to wonder why the gap exists, you probably realized that the San Gabriel River had something to do with it. Just imagine for the moment that you were the San Gabriel River flowing south from Azusa before there was a Whittier Narrows. What would you do when you got to the Whittier Hills?

The question really is, which was there first, the Whittier Hills or the San Gabriel River? In most parts of the country the answer would be the hills. Rivers are mostly young, geologically speaking, and hills are usually older. California, however, just doesn't conform. Geological relationships indicate that the San Gabriel River was here first and that the Whittier Hills were uplifted across its path. Uplift was slow enough so the river was able to cut down fast enough to maintain its course. The Whittier Narrows were cut gradually as the hills rose.

The river anteceded (came before) the hills, and in its course across the Whittier Hills it is an *antecedent stream.*

The Los Angeles Basin has other antecedent streams, such as Los Angeles River across the Elysian Park anticline near Riverside Drive and across Dominguez Hills alongside the Long Beach Freeway, and Santa Ana River across Santa Ana Mountains between Corona and Anaheim and across Newport Mesa near Costa Mesa. Antecedent streams are rather rare birds, and we have a handful right here at home. Be proud of them.

MOJAVE DESERT

People from the leafy greenness of more water-wealthy parts of the world can be unenthusiastic about deserts. That's a shame, because desert country has a charm all its own. Most geologists love the desert, because everything geological is so well exposed; no false eyelashes, no cosmetics, no fancy clothes, just pure plain naked geology with a good coat of tan (rock varnish). A little acquaintance with arid-region geology may help develop your taste for the desert, making travels therein more fun. If you already like the desert, geology can greatly increase your enjoyment of it.

The Mojave is neither one of the largest nor one of the harshest deserts, but geologically it must be one of the more varied. It has just about everything in the way of rocks, landscape, structures, and processes.

Character and Dimensions

A glance at Figure 2–1 shows two things about Mojave Desert. It is the largest of our southern California provinces, and the wedge-shaped western end (Photo 2–8) looks like the bow of a ship sailing into the rest of California. This wedge is outlined by two of California's largest fault zones, the San Andreas on the south and the Garlock on the north. The analogy with a ship sailing west would be reasonable except for one thing—the captain has the engines reversed. Right-lateral slip on the San Andreas and left-lateral slip on the Garlock indicate the Mojave block is moving relatively eastward, not westward.

For convenience, we refer to the western Mojave as that part lying west of the Mojave River between Victorville and Barstow and a line extending northwest from Barstow to Red Rock Canyon (Figure 2–8). The eastern Mojave composes the remainder and much the larger part, extending to the Nevada border and Colorado River.

Topographically, the eastern and western Mojave are distinct. The west consists of great expanses of gentle surface with isolated knobs, buttes, ridges, and local hilly areas. It is a striking landscape when shadowed and highlighted by low-angle morning or evening

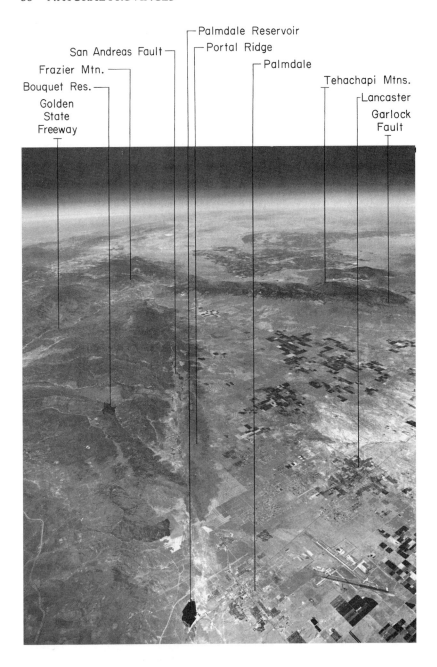

Golden State Freeway
Bouquet Res.
Frazier Mtn.
San Andreas Fault
Palmdale Reservoir
Portal Ridge
Palmdale
Tehachapi Mtns.
Lancaster
Garlock Fault

Photo 2–8.
High-altitude oblique view of western Mojave Desert, looking northwest along trace of San Andreas fault. (Photo taken by U.S. Air Force for U.S. Geological Survey, 064 R-140.)

illumination. Under these conditions isolated knobs look like ancient castles guarding the land. The far western part is exceptionally flat, being an area of deep alluvial fill. Eastward the alluvium thins, and much of the gentle terrain is bedrock reduced to low relief by weathering and erosion. This western region also harbors several good sized dry lakes of which Rosamond, Rogers (Muroc), and Mirage are the best known.

Although smooth and low relative to bordering mountains, the western Mojave is hardly a lowland. Much of it is 2,000 to 2,500 feet above

---

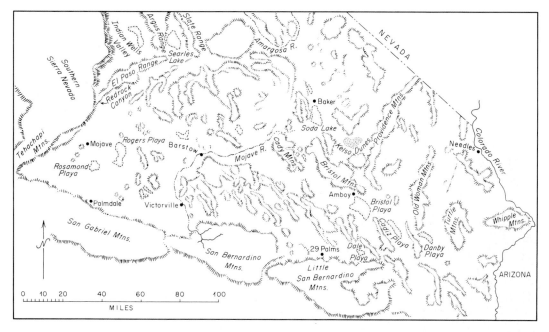

Figure 2–8.
Mojave Desert province.

sea level, which is higher than the top of Palos Verdes Hills (1,480 feet). The knobs and ridges are termed *residual* because they are erosional remnants of once much larger masses. Even so, some rise 1,000 feet above the surroundings, and the top of Soledad Mountain, just west of Mojave town, is 4,183 feet above sea level.

The eastern Mojave also has its share of gentle topography but largely in the form of wide basins and open valleys between mountainous masses. In the southern part, these mountains and valleys have a northwest alignment, like a herd of caterpillars (judging by their shapes in the figures) heading for lunch. In the northern half, the caterpillars are either well fed or haven't heard the dinner bell. Some are headed northeast, some west, some northwest, and some south—confusion reigns.

Furthermore, two large troughs extending east-southeastward are out of phase with any pattern. One trough starts from Barstow and is followed partway by I-40, National Trails Highway 26 (old U.S. Highway 66), and the Santa Fe Railroad. A somewhat smaller trough runs parallel from Victorville to a cul de sac at Dale Dry Lake, 20 miles east of Twentynine Palms (see Figure 2–8). Apple and Lucerne valleys lie in this trough. Some people regard these troughs as possibly the product of streams formerly draining to the Colorado River, but structural influences, both faulting and warping, merit consideration as possible causes, especially for the Barstow-Amboy trough with its string of five young volcanic cinder cones.

Photo 2–9.
View northward to Kelso dunes, highest point 500 feet above camera, sand about 700 feet thick.

The eastern Mojave has its greatest relief farthest east. There, some of its valleys are only 600 to 800 feet above sea level, and higher peaks and ridges attain elevations of 4,500 to nearly 8,000 feet on Clark Mountain (7,929). The north face of the Providence Mountains, as viewed from Kelso Dunes in the late afternoon, is an impressive mountain scarp rising 5,000 feet above Kelso Valley. The dunes are also impressive (Photo 2–9).

Rocks

If someone asked which southern California province would supply the greatest variety of rocks for an avid collector, the answer would surely be, "Mojave Desert." The menu offered equals that of a first-class cafeteria. There are old rocks (late Precambrian), medium-old rocks (Paleozoic), not-so-old rocks (Mesozoic), and young rocks (Cenozoic). All the rock classes— igneous, metamorphic, and sedimentary—are represented in great variety.

The oldest, Precambrian rocks are mostly gneisses and associated metamorphic types.

However, not all metamorphic rocks in the Mojave are necessarily very old. Some of the not-so-old rocks (Mesozoic) are locally highly metamorphosed too. However, Precambrian rocks are invariably strongly metamorphosed. They have been around too long to have escaped severe squeezing, heating, and recrystallization.

Some of the highly metamorphosed rocks of this region are known to be early Precambrian for another reason. They are overlaid by 7,000 feet of only slightly metamorphosed conglomerate, sandstone, shale, and carbonate beds of late Precambrian age. This occurs in the Kingston Range just outside of the northeast boundary of the Mojave Desert as defined here. These unfossiliferous late Precambrian sedimentary rocks are overlaid in turn by 20,000 feet of fossiliferous medium-old (Paleozoic) sedimentary rocks.

Strongly metamorphosed exposures of late Precambrian, Paleozoic, and Mesozoic rocks in some parts of the Mojave commonly form large inclusions within igneous bodies. For example, the marble mined for cement near Victorville, is of the same age as unmetamorphosed limestone in the Paleozoic sequence of the far-eastern Mojave.

Mesozoic metamorphics are exposed mostly in the west-central part of the eastern Mojave and are notable for their high content of volcanic material. This was only one of several intervals of volcanism in the region. However, the most abundant and widespread Mesozoic rocks are coarse-grained granitic intrusives. A generalized geological map of Mojave Desert shows that about 50 percent of the surface is alluvium. Of the remainder, at least 40 percent by conservative estimate is granitic rock, largely of Mesozoic age. These Mesozoic granitic intrusives are 80 to 170 m.y. old.

Granite is normally regarded as a durable rock, and in most places it is. However, in the desert, granite disintegrates so readily that it erodes more easily and rapidly than most metamorphic, volcanic, and hard sedimentary rocks. Hence, some of the lower, smoother parts of the western Mojave are underlaid by coarse-grained granitic intrusive bodies.

Following the Mesozoic igneous activity, the Mojave area was uplifted and stood as a relatively high land mass shedding rock debris to other areas. A thickness of rock amounting to roughly 15,000 feet was removed at this time. By the middle of the young-rock period (Cenozoic) the Mojave region was topographically subdued. Then in about mid-Cenozoic time, warping and faulting created local basins into which rock detritus of local origins was carried. Deposits in some of these basins attained thicknesses of many thousands of feet. Concurrent volcanism created thick piles of lava and contributed large quantities of fragmental volcanic material to the basin fillings. Volcanism has continued at intervals practically to the present.

You may wonder why so many of the rocks in the desert are dark brown or black. Actually, they are of all colors, but a thin coating rich in iron and manganese known as *rock* or *desert varnish* makes their surfaces dark. Varnish forms in different degree on different rocks and is strongly controlled by the environment. Deserts provide optimal conditions for its development and preservation.

Structures

Let's return to our herd of caterpillars in the south. They appear to be crawling northwestward because the country, like a loaf of bread, is sliced by more than 20 major and many minor northwest trending faults. Some of you may recognize a familiar name or two in this list: Mirage Valley, Blake Ranch, Helendale, Muroc, Old Woman Springs, Lenwood, Lockhart, Harper Lake, Camp Rock, Copper Mountain, Calico, Blackwater, Mesquite, Pisgah, Ludlow, and Bullion. These are but some of the faults. The same pattern extends into the western Mojave, but it is less clearly expressed in the subdued topography of that region. Scarps breaking young alluvial surfaces along some of these faults indicate relatively recent displacements. The 1992 Landers earthquake, with a magnitude of 7.5, confirmed their continuing activity. Right-lateral offsets suggest that these faults may be part of the greater San Andreas system.

In the north the caterpillars are confused, crawling in all directions. As you surmise, this is because the faults are of inconsistent trend. Some bearing northwestward are extensions from the southern group, but a good many trend northeast, east-west, and even due north. Garlock fault bounds the Mojave on the north. With an east-west course (see Photo H–1) and left lateral displacement, it represents a very different stress system than the San Andreas. This may be a source of much of the confusion.

Paleomagnetic data from lava flows within blocks between the many faults of the east-central Mojave indicate two episodes of rotation of the blocks by 30° to 50° clockwise.

A north-south line through Baker (see Figure 2–8) roughly marks a boundary in the eastern Mojave between two different types of country. Keep your eyes open if you drive on to Las Vegas. To the west are irregular mountain ridges and masses composed largely of metamorphic, intrusive igneous, and volcanic rocks. To the east are higher, better defined, more linear mountain masses that display thick sections of well-layered Paleozoic sedimentary rocks. The separation between these regions is roughly the trace of the Furnace Creek-Death Valley fault zone (see Figure 1–1). Farther north this is a prominent structure. Here in the Mojave it is less clearly defined, but nevertheless it may be one of the more significant structures of the region.

Resources

Because the Mojave Desert has a varied geology, one might expect a variety of natural resources. Indeed, most of the common metallic elements have been sought and discovered, with significant production of some: gold and silver at Mojave; gold, silver, and tungsten at Johannesburg, Randsburg, and Atolia; silver at Calico. Additionally, iron, copper, lead, zinc, molybdenum, antimony, tin, uranium, and thorium have been produced in smaller amounts. One of the world's richest mines for rare-earth elements is at Mountain Pass (see field trip Segment F).

At present the more valuable products are nonmetallic materials, of which borax is the most notable. About midway between Mojave and Barstow, near Boron, is the greatest borax-producing area the world has ever known, discovered by accident when a homesteader drilled a water well. Here, layers of borax minerals occur within lake beds de-posited in a Cenozoic basin of terrestrial sedimentation. The borax is mined in a huge open pit. The other major nonmetallic product is cement that comes largely from marble bodies in the Victorville area, from Cushenbury Canyon north of Lucerne Valley, and from the Tehachapi Mountains east of Mojave. Volcanic cinders are mined for light-weight aggregate in several places.

Special Features

Dry lakes. Mojave Desert is an area of interior drainage, which means that it does not drain to the sea. The Mojave is the largest river of the province and in terms of flood discharges the second largest stream in southern California. If it had enough water it would flow all the way from San Bernardino Mountains to Death Valley. In fact it did just that, through a string of lakes within the last 11,000 to 15,000 years.

These are some reasons for thinking that at earlier, but still geologically young, times the Mojave River, with possibly a major tributary from the Death Valley country, flowed southeastward through Bristol, Cadiz, Danby, and Chuckwalla valleys (and lakes) to the Colorado River. Related species of fish in some of the present-day landlocked desert basins and features of the landscape support this interpretation of an integrated network of through-flowing streams. Now the drainage has been disintegrated, that is, it has been broken into unconnected bits and pieces. Climatic desiccation, volcanic activity, faulting, warping, and building of alluvial divides (mostly fans) have accomplished this disintegration. In the process many separate basins were created, the floors of which are now occupied by dry lakes—at least they are dry until flooded by occasional heavy rains.

The floors of these lakes are called *playas.* You should make their acquaintance. A playa is one of the flattest natural features on land. One-half inch of water usually suffices to cover many square miles of a playa. If a wind comes up, all the water may blow to one end. Playas are fun to drive or camp on, but keep off of them when they are wet. Even if you don't get stuck, which would serve you right, you disturb their ecology greatly with deep wheel tracks.

Geologists sometimes speak of "wet" playas. This reference is to playas located over underground water at such shallow depths (10 to 20 feet) that moisture can move up small capillary passages to the surface. As it evaporates, this capillary water deposits salts in the soil that fluff up the ground like cookies or a good pie crust. The result is an uneven puffy surface that is poor for driving and into which you sink 2 or 3 inches on foot. Soda Lake south of Baker is a "wet" playa in places, as is Lucerne Dry Lake and many others.

Recent volcanoes. Volcanoes are fascinating as long as they aren't too hot, too active, or too close to you. The Mojave has some beauties. They are perfectly safe even though some were active during the last few thousand years. The most accessible is Pisgah, 35 miles east of Barstow alongside Interstate 40 to Needles. Pisgah volcano is a cinder cone about 250 feet high surrounded by a succession of lava flows, some of them very young. *Cinders* are simply small broken chunks of highly porous volcanic rock. The cinders at Pisgah are mined for lightweight aggregate, and the mining company usually keeps the road to the cone blocked part way in. That, however, need not prevent you from going as far as you can by car, and then walking around on the surface of the lava flows or climbing the cone to look down into its central crater.

Amboy Crater and associated lava flows just west of Amboy are about as fresh. The Bureau of Land Management has recently made Amboy cone more accessible by regrading the entrance road and creating a parking area part way south from National Trails Highway. Foot trails leading to the cone are obvious. Easiest access to the central crater is by way of a trail through the breached western wall of the cone rather than the steep path directly up its northern side. If you really like volcanoes, take the Kel-Baker road southeast from Baker. On its way to Kelso, it passes alongside a Mio-Pleistocene volcanic field of some 52 vents and more than 65 lava flows. More than 30 of the cones and flows are of Pleistocene age, 1.8 m.y. or younger, the youngest about 15,000 years old. Besides flows and cones there are some explosion rings and ash-flow deposits. Both pahoehoe and aa lavas are represented. This is the Cima volcanic field. Mining of cinders from the cones has furnished lightweight aggregates for building many of the gambling palaces of Las Vegas.

The Barstow Formation. Of outstanding interest among land-laid Cenozoic deposits in the Mojave is the Barstow Formation, (see Photo B–1), which consists of about 5,000 feet of sandstone, siltstone, shale, and fine volcanic debris. These beds are well exposed in Rainbow Basin, 10 miles north of Barstow. Interest in the Barstow beds comes primarily from remains of vertebrate mammals living in the area some 15 m.y. ago, including dog-like bears, mastodons, a variety of large and small horses, camels, pronghorn antelope, pigs, dogs and hyena-like dogs, saber-toothed cats, deer, and a variety of small rodents. The country was then more moist, and the abundance of grazing animals suggests extensive grass lands. There were also some palm trees. Rainbow Basin is now a protected state park.

SALTON TROUGH

The heading properly brings to mind the Salton Sea and Imperial Valley. However, viewed from the southeast, one sees that the Salton Trough is basically a landward continuation of the Gulf of California extending northwestward all the way to San Gorgonio Pass (Figure 2–9). It thus includes Coachella Valley and Palm Springs. In addition, the northeastern boundary is drawn to encompass mountain ranges that don't fit comfortably into either the Transverse Ranges or Basin Range provinces. These mountains are the Orocopia, Chocolate, Chuckwalla, Palo Verde, and Cargo Muchacho, intriguing names and geologically interesting areas.

Salton Trough and its immediate environs provide a menu of considerable variety. Items for consideration include large faults slicing across the area, recent volcanic knobs at the southeast end of Salton Sea, hot brine wells, sand dunes, Colorado River flooding, and the past history of the Salton Sea.

Character and Dimensions

Salton Trough embraces the largest area of dry land below sea level in the Western Hemisphere, something over 2,000 square miles. Death Valley is a little deeper, measuring –282 feet compared to –273 feet, but the below-sea-level area in Death Valley is much smaller.

Salton Trough extends 140 miles northwestward from the Gulf of California to San Gorgonio Pass. It is only a few miles wide at the northwest end but attains a maximum width of roughly 70 miles at the Mexican border. Geologically this is a subsurface structural feature of impressive vertical dimensions. East of Mecca, on the northeast margin, the buried

bedrock basement abruptly rises 12,000 feet. Under Imperial Valley the depth of unconsolidated fill exceeds 20,000 feet.

The trough is bordered by rugged mountains on both sides, those to the west being higher and more massive. The western margin southward from the end of Santa Rosa Mountains is made irregular by such major embayments as Clark and Borrego valleys, and forelying hills such as the Superstitions, the result of complex geological structures.

Rocks

The bordering mountain ranges consist largely of metamorphic and igneous rocks, some old (late Precambrian) and some not so old (Mesozoic). The units recognized include the Chuckwalla complex, Orocopia schist, Pinto gneiss, and intrusive granitic rocks like those of the southern California batholith in the Peninsular Ranges. Of special interest are young plutonic bodies only 23 to 31 m.y. old.

The trough is filled with young (Cenozoic) sedimentary deposits, largely of land-laid origin but partly marine. The contents of these land-laid beds change rapidly from place to place and coarse gravels are commonly intermixed with layers of finer sandstone, siltstone, and mudstone. Locally some volcanic rocks are included. The oldest Cenozoic beds probably are younger than 20 m.y. The filling is thickest in the southern part of the trough, where a well 12,000 feet deep failed to penetrate the section, and geophysical explorations indicate about 21,000 feet of soft sediments above the basement.

Within the Cenozoic units the Imperial and Split Mountain formations are unusual in being largely of marine origin. The Imperial is composed

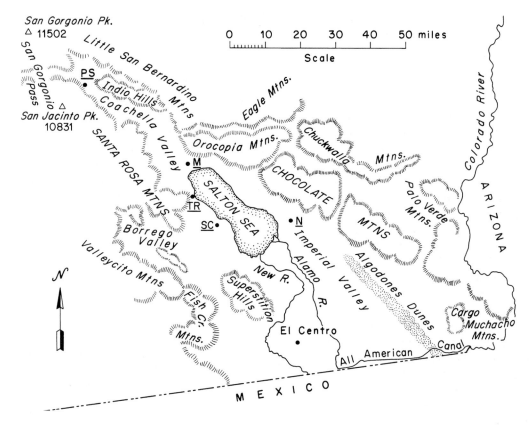

Figure 2–9.
Salton Trough province; M-Mecca, N-Niland, PS-Palm Springs, SC-Salton City, TR-Travertine Rock.

mostly of brownish sandstone and mudstone, lo-cally rich in marine fossils (snails, clams, oysters, and corals) and is preserved in isolated patches from the Mexican border almost to San Gorgonio Pass. The fossils are not normal Pacific Coast types. They are tropical and apparently migrated up the Gulf of California. The Imperial Forma-tion is estimated to be 5 to 10 m.y. old, and the Split Mountain is somewhat older.

At the northern tip of Fish Creek Mountains along the west side of southwest Imperial Val-ley, a deposit of lake beds rich in gypsum un-derlies the Imperial Formation. It is extensively mined and the gypsum goes by narrow-gage railroad 25 miles to Plaster City, where it is processed into plaster and plaster-board. Other unusual rocks outcrop in the vol-canic knobs at the southeast end of Salton Sea. Here is *obsidian* (natural volcanic glass) and *pumice* (frothy volcanic glass), so light it floats on water. The floor of Imperial Valley is thinly veneered by fine grained lake deposits, locally rich in snail and clam shells. These were laid down when expanded stages of the Salton Sea formerly covered more of the valley.

Structure

Southern California's three big right-lateral fault zones, the San Andreas, San Jacinto, and Elsinore, traverse this province. In fact, there are so many northwest trending faults in the area that it's hard to know to which system individual fractures belong. This is partly because the San Jacinto fault largely, and the Elsinore partly, splay into several branches as they approach Imperial Valley.

Representatives of the San Andreas system can be traced along the northeast side of Salton Trough to the Algodones Dunes (Sand Hills). This dune belt is aligned along a fault that is probably the southern continuation of the San Andreas. The San Jacinto separates into a complex of faults in Borrego Desert and southward. Among these fractures, the Imperial fault is particularly noteworthy. It has generated three sizable earthquakes within modern historical times: 1915, 1940, and 1979. This fault can be traced for many miles by a 10 to 15 foot scarp extending northwest about 3.5 miles east of the town of Imperial. If one looks carefully at the alignment of power line poles, highways, railroads, the All American Canal, and even the Mexican border where crossed by the Imperial fault, they all show the effects of right-lateral displacement. The border was offset nearly 15 feet in the 1940 earthquake. Branches of the Elsinore fault help define the western border of Salton Trough and continue into Mexico.

Some Imperial Valley faults are currently experiencing slow continuous slippage. This produces cracks in highways and concrete irrigation ditches where crossed by a fault. Highly precise resurveys of Coast and Geodetic markers demonstrate current differential slip of opposing blocks along the Imperial fault of 1 to 2 inches a year. Thanks to all this restlessness, Imperial Valley is one of the most seismically active parts of California, with more than 30 moderate to large earthquakes within the twentieth century. Total regional strain in the area accumulates at between 2 and 3 inches per year—a lot of strain.

Folds have formed along the margins of Salton Trough where the young (Cenozoic) sedimentary filling has been uplifted. They can be seen off the southeast end of Santa Rosa Mountains and south into Superstition Hills. Folds are also found along the northeast margin of the trough in the Mecca and Indio hills. A ride up Painted Canyon east of Mecca provides a good view of severely deformed Cenozoic beds. By watching the inclination of bedding you can recognize several folds. One or two of the anticlines have old metamorphic rocks squeezed into their cores.

Geologists have concluded that the Gulf of California was created within the last few million years by plate tectonic movements that are ripping Baja California away from mainland Mexico and are carrying it to the west-northwest. The force is attributed to the same sort of deep convective currents within the earth that cause spreading of the sea floor outward from mid-oceanic ridges. Because Salton Trough is simply an extension of the Gulf of California, it too has presumably been formed by the same process, but it is in a less advanced stage of development. If the process continues, San Diego may eventually find itself separated from Arizona by an arm of the sea.

Resources

Aside from plaster products made from the Fish Creek Mountains' gypsum, dry ice (solid carbon dioxide) was formerly produced near Niland, but the rising Salton Sea flooded the carbon dioxide gas wells and put them out of commission about 1957.

Gold was mined in the Cargo Muchacho Mountains as early as 1781, and off and on in later years. Currently, the Mesquite mine at the south end of the Chocolate Mountains is one of the most productive gold mines in California. Its huge leach piles are just off Highway 78, about 6 miles northeast of Glamis. The low-grade ore comes from nearby alluvially buried areas of bedrock, mined in several open pits. The Picacho mine, 15 miles southeast, is similar but less accessible.

Major efforts since 1957 to exploit the power and mineral resources of the hot brines deep under Imperial Valley have only recently become successful, as explained later in the chapter. The greatest and most successfully employed resource of the region is the rich soil on which the farms and orchards of Imperial and Coachella valleys are located.

Special Features

Salton Sea. Between 1900 and 1905 the floor of Salton Trough was dry, although in the 1880s and again in 1891 overflows from the Colorado River had created lakes a few feet deep in its lowest part. In 1905 the main flow of the Colorado River was accidentally diverted by way of irrigation canals into the Salton Trough. This was the beginning of the present Salton Sea. Only the herculean efforts of Imperial Irrigation District and Southern Pacific Company restored the river to its normal channel by 1907. By then flood waters flowing down the channels of the Alamo and New rivers had created a lake nearly 80 feet deep, covering over 400 square miles, with a water level and shoreline at –195 feet. Once inflow was shut off, the lake began to shrink because of high evaporation rates, and by 1925 it had stabilized at –250 feet. The lake fluctuated around that level until 1935 when it began to rise again. It has risen steadily ever since to the 1970 level of –230 feet. Completion of the All American Canal introduced more water into Imperial and Coachella valleys and was the cause of this rise.

About 6 feet of water are evaporated from the surface of Salton Sea each year, making the water saltier, so by 1980 it was more saline than the ocean. As the size of the sea increases, so does the area from which evaporation occurs. Calculations suggested that eventually a balance would be struck between inflowing water and evaporation loss. This happened about 1981 at a level near –227 feet. All lands bordering Salton Sea below –220 feet had judiciously been withdrawn from public occupation in order to prevent economic loss. The rising level had made islands out of some of the volcanic knobs at the southeast end of Salton Sea and submerged the former hot springs, steam jets, mud pots, and mud volcanoes of the Niland area. Boat landings also suffered.

Because human beings accidentally created a good-sized lake during 1905 to 1907, the question arises, "Might not nature have done the same thing, and perhaps on a larger scale, at some earlier time?" Around the edges of Salton Trough at about 40 feet above sea level are prominent shoreline features of lakes well over 300 feet deep that occupied the trough repeatedly over the last few thousand years. The succession of lake-bed deposits shows that water rose to the +40 feet level, at least four times in the past 2,000 years, most recently about 400 years ago. This water body goes by the names Lake Cahuilla or Lake Le Conte. It appears to have been held at the +40 foot level by the crest of the Colorado River delta separating Salton Trough from the Gulf of California.

Along the highway on the east side of Salton Sea a prominent Lake Cahuilla beach ridge of gravel can be seen back of Hot Mineral Spa. Along the west side, a superb Lake Cahuilla shoreline is cut into the mountain front in the vicinity of Travertine Rock (Photo 2–10). Rocks above the shoreline are pale because wave cutting formed a cliff and destroyed the rock varnish still preserved at higher levels. Rocks below the shoreline are coated with dark travertine. The *travertine* (a porous, irregular deposit of calcium carbonate) of Travertine Rock (see Figure 2–9) alongside Highway 86 was deposited below the surface of Lake Cahuilla. This encrustation is as much as 30 inches thick.

Photo 2–10.
Horizontal shoreline marks formed at base of Santa Rosa Mountains near Valerie Jean by an expanded Salton Sea (Lake Cahuilla phase.) (Photo by John S. Shelton, 3600.)

There are scattered remnants of shorelines of still older and larger lakes around the margins of the trough. Some of these bodies date back thousands of years, and judging from the shell life living along their shores, one or two had connections with the gulf. This means that the oceans either stood higher then or that the shoreline features have subsequently been uplifted. The latter seems more likely.

Sand dunes. Many of you have already seen the Algodones Dunes (Sand Hills) along the southeast edge of Salton Trough without leaving your home. They are favored by film and television companies filming adventures purporting to occur in the Sahara. This area is also much used by dune-buggy enthusiasts.

Algodones Dunes comprise a chain 45 miles long in a north-northwesterly direction and 4 to 8 miles wide. Sand within this belt attains a maximum thickness of nearly 400 feet. One of the impressive features of this chain is a series of subequally spaced intradune flats, seen particularly well in the southern third of the belt from the air (Photo 2–11). These flats are egg shaped areas within the dunes, from which sand has been removed by the wind down to the underlying base of gravel. The All American Canal and U.S. Highway 80 cross the dune chain by means of a small intradune flat (Photo 2–12). The flats are dynamic, moving south-southeastward at 6 to 12 inches per year. This movement will eventually bring a huge dune mass right up to the All American Canal. However, it looks as though that unhappy day is about 400 to 500 years in the future. By that time the canal will have paid for itself several times over, and we can afford to construct a new canal.

On the west side of Salton Sea in the vicinity of the old Salton Sea Test Base, seven miles south of Salton City, is a field of barchan dunes

Photo 2–11.
Intradune flats in southern Algodones Dunes viewed northward. U.S. Highway 80 and All American canal traverse second flat. Foreground flat and associated dunes much used in movies. (Photo by John S. Shelton, 1816.)

Photo 2–12.
Oblique air view west over intradune flat in Algodones chain, traversed by All American canal and highway, trees mark rest area. (Spence air photo E-8589, 04/07/38, courtesy of the Department of Geography, University of California, Los Angeles.)

Photo 2–13.
Barchan dunes west of Salton Sea south of Salton City, viewed north-northwest. (Photo by John S. Shelton, 1807.)

(Photo 2–13). *Barchans* are individual dunes of symmetrical crescentic shape. They develop best in areas where the wind blows mostly from one direction, where the sand supply is not too abundant, and the ground is smooth. The crescents open downwind. Little barchans move more rapidly than large ones, so sooner or later they catch up with larger dunes that gobble them up, making the large barchan even bigger and slower.

Hot brine wells. Many countries, but chiefly the United States, Philippines, Iceland, Mexico, Italy, and Japan, have used the interior heat of the earth to generate electrical power, the United States being far in the lead in terms of current production, thanks to the operations at the Geysers in Sonoma County, central California. Iceland is unquestionably the leader in using geothermal waters for space heating.

The discovery of hot brines in a deep well near the southern tip of the Salton Sea in 1957 excited considerable interest and led to the drilling of hundreds of wells, many 5,000 to 8,000 feet, and some over 13,000 feet deep, in parts of the Salton Trough. As a result, nine potential geothermal areas were defined, of which the Salton Sea, North Brawley, Heber, East Mesa, and Cerro Prieto (30 miles into Mexico) are considered major and most promising. Of these, Sierra Prieto is the longest one producing a significant amount of power for domestic use. The amount of hot brine estimated to be available at temperatures of 400° to 600°F is huge, possibly enough to serve 4 million people for thousands of years.

But there are problems. The brines are hard to handle because they are corrosive. Thus, they clog pipes and are disposed of safely only by being pumped back down deep recharge drill holes. Problems are slowly being solved, and 35 years after discovery the brines at long last are coming into their own. The successful tapping of this huge resource is an event to be celebrated. In early 1993, the power coming online was estimated to be close to 400 megawatts. That is sufficient to support a city with hundreds of thousands of citizens. The problem of deposits choking transmission pipes was solved by running the primordial brine through precipitators before circulating it into the power-generating systems. The precipitate, mostly silica, is then disposed of in various ways, some useful, and only clear, hot water is used for power generation.

It has been speculated that substances in the brines, especially potassium, lithium, and a host of metals—gold, silver, copper, nickel, chromium, lead, zinc and others—have a value possibly surpassing the potential for electrical power generation.

Experimental power plants can be seen close up between Heber and Calexico south of Interstate 8 and west of Highway 111 and at least five large plants are now (1993) operating at the south end of Salton Sea. The brines are a witches' brew, and it requires a sorcerer to tame them.

Wind farms. Anyone visiting Coachella Valley will be astounded by assemblages of thousands of windmills of various types on top of high towers. They are generating electrical power. San Gorgonio Pass and that part of Coachella Valley immediately east have long been known to be one of the most consistently windy spots in California, with strong unidirectional west winds prevailing. Although wind is advocated as an inexpensive, nonpolluting, renewable source of power, its current benefit is only that it's renewable. Despite tax credits, windmills so far have proved to be more expensive than power from other sources. The capital investment is large, exceeding $600 million in Coachella Valley alone; and the noise, disfigurement of the landscape, and danger from flying propeller blades on nearby highways constitute a special type of pollution.

In 1985 the city of Palm Springs brought suit against the partnerships that installed the windmills and the Bureau of Land Management, which granted right of ways on public domains, seeking some limits on the number and location of windmills. The city claimed gross violations of the previous impact statement with respect to those items. Growth in the number of windmills seems to have stabilized at about 4,000 by 1993, probably mainly because the federal tax credit expired in 1985 and the Palm Springs suit (settled out of court).

SOUTHERN COAST RANGES

We treat the southern Coast Ranges briefly, but only because they are peripheral to our geographical focus. Southern Coast Ranges country is delightful, and the geology is interesting.

If you look at the western sector of the Transverse Ranges with its thick sedimentary section

and many faults and folds, rotate it so the grain runs northwest, and you have the basic characteristics of the southern Coast Ranges. Add a rugged coastline (Photo 2–14) and some of the loveliest coastal dunes of the Pacific shore, south of Pismo (Photo 2–15), and you begin to sense the essence of the country. Morro Rock is

Photo 2–14.
California coast west of San Luis Obispo 5 miles north of Point Estero, a wave eroded shoreline.
(Spence air photo E-13256, 09/19/47, courtesy of Department of Geography, University of California,
Los Angeles.)

Photo 2–15.
Coastal dunes near Pismo, part of the Nipomo dunes complex. On-shore winds transport sand inland from the beach. (Photo by John S. Shelton, 4393.)

a volcanic plug, the last one in a line of 13 extending northwest from south of San Luis Obispo (Figure 2–10).

The sedimentary section in the southern Coast Ranges is thick, and every geological epoch from Cretaceous to Holocene (*see* Appendix A) is represented by layers of sandstone, conglomerate, or shale. Deformation has occurred intermittently over tens of millions of years, and the oldest beds are the most severely folded and faulted. Movement has continued right up to the present, so even young beds are deformed.

The region boasts a host of faults, two of which that bear northwesterly merit particular notice. One is our old friend, the San Andreas, and the other is a parallel fault about 30 miles to the west, the Nacimiento (see Figure 1–1). No place

is better than Carrizo Plain, near the eastern edge of the southern Coast Ranges (see Figure 2–10), in which to see the impact of the San Andreas on the land. Commuter flights between San Francisco and Los Angeles frequently travel right along the San Andreas in this area, so keep your eyes peeled. Owing to gentle relief, scanty vegetation, and the transverse crossing of drainage courses by the fault, the evidence of displacement is spectacular (Photo 2–16). The fault makes a fresh scar across the land. Within the fault zone are small depressions, marshes, and wet spots called *sags,* or *sag ponds* when wet. They mark places where a slice within the zone has sunk. Uplifted slices compose fault ridges. Scarps and elongated narrow troughs are abundant. Clearly evident is the lateral offset of stream courses, showing right-lateral displacement along the most recent line of movement.

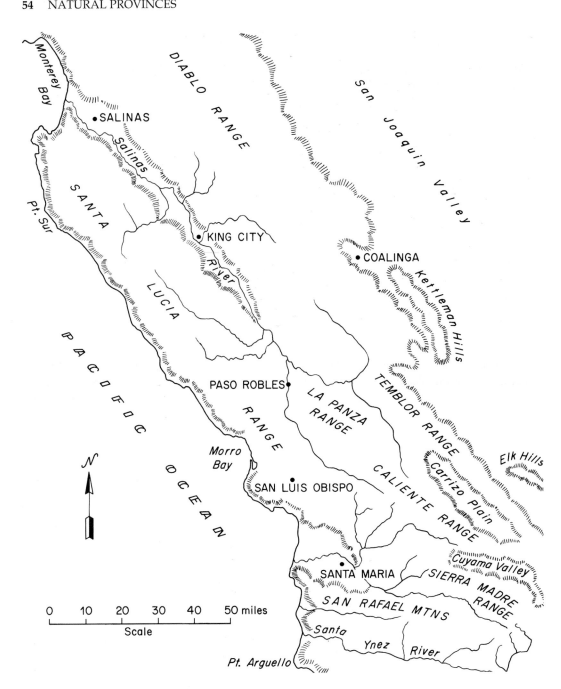

Figure 2–10.
Southern Coast Range province.

Photo 2–16.
Trace of most recent break within San Andreas fault zone in Carrizo Plain, viewed south-southeastward. (Photo by John S. Shelton, 510.)

The Nacimiento fault is regarded as relatively old but not necessarily dead. Like old scars on the human body, old faults on Mother Earth are partly obscured by healing. The Nacimiento is a complex zone that extends from Cuyama Valley northwest at least to Point Sur, and it probably continues offshore into the submerged continental borderland. It separates two distinct types of basement rocks. To the northeast are granitic rocks, averaging about 90 m.y. old, intruding an older metamorphic complex of gneiss, schists, quartzite, and marble—the Sur Series. To the southwest is a wholly different series of peculiar, metamorphosed, dark sandstone, shale, conglomerate, volcanic, chert, and limestone layers known as the Franciscan Series. Associated are some iron-magnesium rich intrusive masses, altered to a slippery soapstone, known as *serpentinite*. Remember that the Newport-Inglewood fault zone in the Los Angeles basin separated two distinct types of basement rocks. The Catalina schist of the southwestern block there is presumably related to the Franciscan Series. Total displacement on the Nacimiento is not known, but it is judged to be considerable. Some has been lateral and some, particularly of younger date, has been vertical.

An unusual feature in parts of the southern Coast Ranges is elongate bodies of serpentinite that have been intruded in a cold state into the core of many upfolds (anticlines). Serpentinite is so ductile and slippery that it moves easily under stress and forces its way into adjoining rocks. The core of an upfold is an ideal place for such an intrusion to occur.

The rock formation most characteristic of the Southern Coast Ranges is the Monterey. Its light-colored shales contain considerable phosphate and are particularly distinguished by their high content of silica in the form of diatoms and volcanic ash. An unusual resource of the Coast Ranges is mercury, in the Franciscan Formation. The coastline of this province features some spectacular sequences of emerged marine terraces.

THE GREAT VALLEY

The Great Valley of California lying between the lofty Sierra Nevada and the lower Coast Ranges is a true heartland. It actually consists of two valleys, the Sacramento in the north and San Joaquin in the south, each drained by an axial stream of corresponding name. Although these two sectors are part of a continuous feature, they display some differences in geology and resources. For example, the San Joaquin has produced more than $10 billion worth of oil and gas, but only gas is obtained in the Sacramento sector. This contrast arises primarily from the differences in the nature of the underlying sedimentary deposits.

Topographically the Great Valley is a smooth alluvial plain a few hundred feet above sea level. Geologically it is a structural trough with a northwest length of 450 miles and a width of 50 miles. In terms of depth to basement rock it is strongly asymmetrical, shallow on the east and deepest to the west. The wedge of sedimentary fill within the trough, accumulated over the last 15 to 20 m.y., has a cumulative thickness in excess of 60,000 feet.

This book is concerned only with the San Joaquin Valley, and specifically with that part south of Fresno (Figure 2–11). The two largest streams debouching into this section, the Kings and Kern rivers, have built huge alluvial fans of such gentle slope that you cross them almost without recognition. These fans block off parts of the valley creating shallow basins, Buena Vista on the south and Tulare on the north (see Figure 2–11). Lakes or marshes occupied these basins until river waters were diverted into irrigation ditches and the land was drained for cultivation, much to the chagrin of duck hunters.

In former high-water years Buena Vista Lake occasionally overflowed north to Tulare Lake, but the latter apparently seldom, if ever, topped the alluvial divide into San Joaquin River. These lakes and the north-flowing San Joaquin River hug the west side of the valley. They have been pushed there by large deposits of rock debris swept down by powerful Sierra Nevada streams only feebly opposed by small creeks from the relatively dry Coast Ranges.

Although the Great Valley has been the site of sedimentary deposition for nearly 145 m.y., it assumed its trough-like form only during the last 60 m.y. This coincided with initial uplift of the Coast Ranges. Westward tilting of the Sierra Nevada-San Joaquin block subsequently deepened the San Joaquin sector, making it a principal site of marine deposition in the last 10 to 20 m.y. The scene must have been striking when a great inland sea filled the San Joaquin Valley and lapped against the west base of the Sierra Nevada or dashed its waves on a narrow coastal plain at its foot. Eventually this inland sea filled with sediment, and marine deposition was succeeded by terrestrial accumulation. The seas lingered longest at the southern end of San Joaquin Valley.

Although the present alluvial floor is nearly featureless, the underlying sedimentary filling has been folded and faulted, especially on the west side. Some of these structures are buried, but others are exposed along the west margin from Coalinga south. Here outlying hills mark the site of a succession of anticlines, of which Kettleman Hills is the most famous. Oil and gas have accumulated in many of these structures. The folds are usually broken by faults, and one near McKittrick leaked oil to the surface creating tar pools in which animals and birds were trapped, just as at Rancho La Brea. The McKittrick fauna was not as rich and varied, but it was still impressive, including such animals as saber-toothed tigers, big lions, lynx, dogs, the big dire wolf, skunks, bears, ground sloths,

Figure 2–11.
Great Valley province.

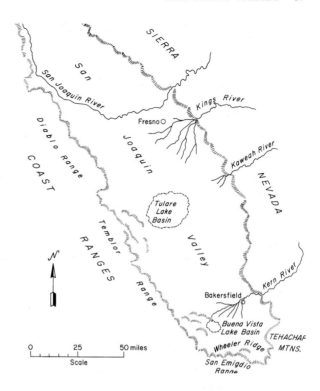

camels, horses, bison, deer, antelope, elephants, mastodons, and lots of rodents. The abundant birds included grebes, herons, bitterns, storks, ibises, swans, ducks, geese, vultures, kites, hawks, falcons, partridge, quail, and cranes. The area was not as dry as at present.

At the southern end of San Joaquin Valley the structures take on a nearly east-west trend, an example being Wheeler Ridge (Photo 2–17). This striking anticlinal fold of recent date, is seen just west of the Golden State Freeway as it leaves the foot of Grapevine Grade northbound. The folding here is so recent that relatively unconsolidated sands and gravels are involved. These deposits are currently being worked for sand and aggregate in a large pit near the eastern end of the ridge.

Wheeler Ridge anticline is asymmetrical, steepest on the north side, with a thrust fault along its north base. White Wolf fault, which caused

the destructive 1952 Arvin-Tehachapi earthquake, as traced west-southwest, passes under Wheeler Ridge. The *epicenter* of the earthquake (spot on the ground beneath which slip first occurred) was just south of the ridge.

Wheeler Ridge is best seen traveling south on I-5 and Highway 99, but northbound travelers can look back to view it in profile. Such a view shows that the crest is notched by several prominent topographic saddles or gaps. These were cut across the growing anticlinal fold by streams formerly flowing north out of San Emigdio Range to the south. Three such gaps can be seen near the east end of the ridge. The largest is no longer occupied by a stream, because it could not cut down fast enough to keep pace with the uplifting. The stream was turned aside, and its abandoned notch is called a *wind gap*. Daniel Boone built the Wilderness Road to Kentucky through Cumberland Gap, a wind gap in the Appalachian Mountains

Photo 2–17.
Wheeler Ridge west of Golden State Freeway beyond foot of Grapevine Grade. A growing anticline with a wind gap (center) and a water gap (east end), as viewed north. (Photo by John S. Shelton, 496.)

produced in a different way. Our wind gap is used by the California Aqueduct as a means of crossing Wheeler Ridge. If the day is clear, you should be able to see the four large pipes rising into the gap from the pumping station at its north base.

Wheeler Ridge is still growing upward at an average rate of about 0.06 inches (1.4 mm) per year. That seems slow to us, but at that rate it took only 200,000 years to create Wheeler Ridge. There are some indications that a brand new anticline is starting to form under the farms out on the valley floor north of Wheeler Ridge. This is not the only current tectonic activity in the area, a thrust fault in Buena Vista Hills creeps about 1 inch per year, and Kern Front fault near Bakersfield slips about 0.5 inch each year.

An ongoing problem in San Joaquin Valley is land subsidence amounting to a good many feet, in one instance 28 feet. This creates havoc with irrigation canals and is not welcomed by other operations. It has several causes, the principal ones being soil compaction through irrigation and withdrawal of water and oil from underground by pumping. Compaction caused by ground wetting was a problem for the California Aqueduct project. Prospective aqueduct channel routes had to be soaked with water before being lined with concrete, because subsidence of as much as 15 feet occurred in 10 to 18 months when the site was wetted.

SOUTHERN SIERRA NEVADA

Fine furniture, rugs, and drapes tastefully arranged can make almost any home attractive, but it helps to start with a good house. The Sierra Nevada are beautifully decorated with streams, lakes, meadows, and forests, and the air conditioning is tops. All this would not be nearly so attractive were it not associated with a spectacular bedrock landscape. This, California's largest mountain range, is an asymmetrical block, rising steeply from a major zone of

Photo 2–18.
East-face scarp of Sierra Nevada west of Lone Pine. Alabama Hills in foreground, Whitney Portal road on scarp face to right, Lone Pine Peak left center.

faults at its east base (Photo 2–18) and sloping gently westward. Great masses of relatively homogeneous, coarse-grained granitic rock and deep sculpturing by running water and flowing ice are the elements that have given the Sierra their special character. We are here concerned only with the south part and principally with the eastern side. A more complete treatment of the Sierra, specifically of the Yosemite-Tioga region, will be found in John Harbaugh's Kendall/Hunt companion book, *Geology Field Guide to Northern California.*

Character and Dimensions

The Sierra is more than 400 miles long and as much as 80 miles across. Other North American ranges are higher, but not in conterminous United States. The difference in elevation be-

tween the highest and lowest points within the range is nearly 14,000 feet. Sequoia Park containing the High Divide, a lofty ridge within the Sierra, has the greatest topographic relief, 13,500 feet, of all our national parks outside Alaska.

The topographic asymmetry of the Sierra is extreme. The crest, in places, lies within 4 to 6 miles of the eastern base and 60 to 70 miles from the western foot (Figure 2–12). Consequently, streams flowing eastward are short and steep, whereas those flowing west are 10 times longer, considerably larger, and traverse a gentler slope. Incidentally, the zone of maximum precipitation in the Sierra is far down the western flank, at elevations between 7,000 and 8,000 feet, not at the crest.

A few Sierra streams flow north or south roughly parallel to the range axis. An example is Kern River (Photo 2–19) which has eroded

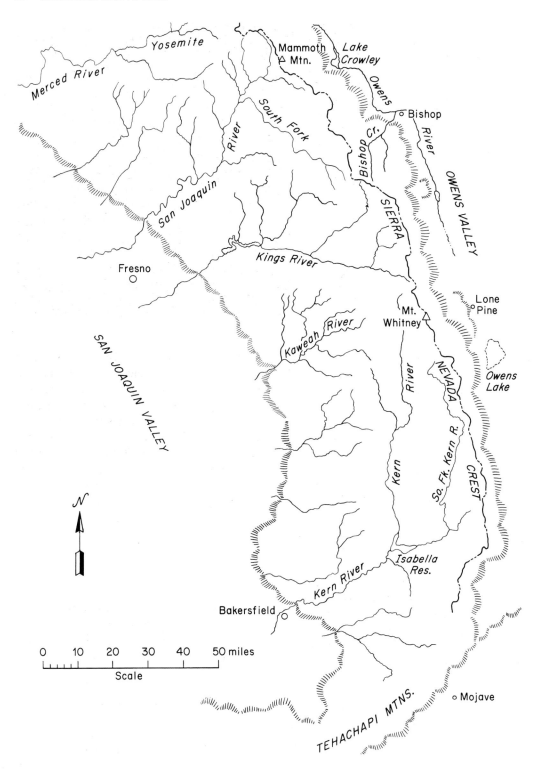

Figure 2–12.
Southern Sierra Nevada province.

Photo 2–19.
Low altitude, oblique view south, of Kern Canyon from near mouth of Whitney Creek. U-shaped glaciated canyon 2,000 feet deep bordered by broad bench remnants of more open older canyon stage. (U.S. Geological Survey air photo, GS-OAL-1–73.)

its valley along the fractured rock of a north-south fault. The upper part of the Middle Fork of San Joaquin River, in the Devil's Postpile region, flows south along a line determined by banded structures in metamorphic rocks, and Bishop Creek flows north along a fault or hinge line. The direction of flow of Sierra streams is important for this reason. The range has been uplifted step by step, growing higher in the eastern part and tilting ever more steeply westward. Imagine that you are a stream flowing down the long western slope. Each uplift of the range steepens your bed, giving you greater energy to cut deeper. Now imagine that you are a Sierra stream flowing north or south. You experience no increase of slope because of tilting, and furthermore, if you are tributary to a west-flowing stream, you will be left "hanging" as the main stream cuts down. This is just one of several possible causes for the many hanging tributary streams in the Sierra with waterfalls at their mouths.

In parts of the Sierra are extensive areas of subdued upland terrain (see Photo M–10) across which travel is easy. The weathering characteristics of homogeneous granitic rocks have a lot to do with the smoothness of these upland flats. In places, however, remnants of such subdued surfaces make prominent benches along the sides of deep canyons (see Photo 2–19), suggesting that at some earlier time these streams flowed in wide, open valleys before cutting down into their present narrow canyons. Thus, at least some of these benches and areas of subdued relief are judged to be the product of extended periods of weathering and erosion occurring between intermittent uplifts of the range.

Rocks

Sixty percent of the exposed rocks of the Sierra Nevada are coarse-grained intrusives of several varieties. They constitute the Sierra Nevada batholith, which actually consists of at least 100 separate *plutonic* bodies ranging from 90 to 200 m.y. old. The rock in some of these bodies is unusually massive with only a few widely spaced cracks (joints). These types of rocks compose the monolithic domes of the Sierra, such as the well-known features in Yosemite.

Because batholithic rocks are intrusive, something older had to be intruded by them. These older rocks are bodies of metamorphics seen along the east side of the range mostly north from Bishop and low on the western slope. They occur largely as dividing partitions (*septa*) between individual intrusive bodies or as *pendants* within them. These metamorphics have been formed from great thicknesses of Paleozoic sedimentary and Mesozoic sedimentary and volcanic rocks. Long ago these rocks were tightly compressed into folds trending roughly northwest. Metamorphism occurred both during this deformation and during the subsequent intrusion of igneous bodies. The usual black and brown colors of these rocks stand out against the lighter shades of the igneous bodies. Any time you see steeply tilted, laminated rocks with highly variegated colors in the Sierra, be suspicious that they are metamorphic. Many of the metal ore deposits of the range lie along contacts between igneous and metamorphic rocks.

Cenozoic volcanics constitute the third principal group of Sierran rocks. Before they were extruded, however, the Sierra region had been strongly uplifted and subjected to extensive erosion that removed an estimated 10 miles of rock, exposing the batholith upon which the volcanics rest. Volcanic rocks occur in greatest amount, variety, and thickness on the west slope of the range from Sonora Pass north. In the southern Sierra, young volcanics occur only in widely scattered spots. On the west slope they appear principally as remnants perched on stream divides in the San Joaquin and Kern river drainages. Some recent flows obstruct the canyon of Golden Trout Creek, and young volcanic cones lie in the middle of the Kern River drainage. Columnar joints are beautifully formed at Devil's Postpile on the Middle Fork of San Joaquin River (Photo 2–20). Along the east base of the Sierras, volcanic cones, flows, and obsidian domes appear in increasing abundance northward from Little

Photo 2–20.
Hexagonal columnar joint block in talus pile at Devil's Postpile, Middle Fork San Joaquin River, Sierra Nevada.

Lake to the Nevada border (see Photo M–2). The Mammoth trip guide will have more to say about these particular volcanics.

The youngest deposits of the range, aside from some of the obsidian domes and flows of the Inyo and Mono craters and the associated widespread pumice, are the famed gold-bearing stream gravels of the west slope and the glacial moraines of both the east and west-slope canyons.

Structure

The Sierra Nevada is usually described as an uplifted fault block tilted west, but this may be an oversimplification. There are reasons for thinking that much of the faulting responsible for the high east-face scarp of the Sierra has involved down-dropping of Owens Valley more than uplift of the mountains. One model currently entertained is that an area embracing the Sierra Nevada, Owens Valley, and country farther east, was warped up into a huge arch attaining its highest elevation about 3 to 4 m.y. ago. Then Owens Valley started to drop by movement along faults at the east base of the Sierra and the west base of Inyo Mountains. This occurred during the last 3 million years with a significant amount of the displacement taking place in the last 700,000 years. Indeed, fault scarplets formed as recently as 1872 across alluvial fans and young volcanics in Owens Valley (see Photos N–5 and O–1) indicate that the process continues. Faulting and warping have also occurred within the Sierra block and along the west base from the San Joaquin River south to the mouth of the Kern.

You may be concerned about what holds the Sierra so majestically above its surroundings. Rocks of the earth's crust are simply not strong enough to support such a massive load. When vibrations generated by earthquakes centered somewhere east of the Sierra arrive at a recording station west of the mountains, it is clear that they have dawdled along the way. Geologists and geophysicists have speculated that what slows them down is a gigantic Sierra root, something like a root to a wisdom tooth. The Sierra may resemble a great iceberg floating in a sea of rock. The force that buoys them up is the displacement produced by the root, in the same way that the part of a boat below water is what keeps it afloat. Rocks composing the root are assumed to be similar to the rocks exposed at the surface. They displace heavier, denser materials and transmit earthquake waves more slowly, thus retarding the earthquake waves that pass through the root. More recently, on the basis of geothermal observations, a contrary speculation has been advanced that the

Sierra are buoyed up by rising and expanding hot rock deep within the crust. Hot rock would also be capable of slowing down earthquake waves, if the waves went deep enough.

Resources

The Sierra are most famous for gold that occurs in placer gravels on the west slope and in the quartz veins of a metamorphic belt near the west base—the *Mother Lode*. They have also produced silver, copper, lead, zinc, chromium, and a significant amount of tungsten. Nonmetallic materials include building stones, limestone, and some barite, a barium sulphate mineral used as a source of barium but also valued for its heavy weight.

Special Features

The Sierra Nevada are rich in special features, but we select for consideration just one aspect of its scenic development: the glaciation. Flowing streams of ice have played a major role in developing scenic characteristics of the High Sierra. Glaciers first formed in these mountains at least a million years ago and have been intermittently active until about 10,000 years ago. The small glaciers currently clinging to shaded sides of high peaks are just babies in their erosional, transportational, and depositional power compared to the giants of the ice ages. Some ice age valley glaciers of the Sierra were 40 miles long and thousands of feet thick.

We tend to think of glaciers too much as scrapers, when in actual fact they are often more effective as excavators. Glaciers do wear down rocks by grinding, producing the eye-catching, scratched, smoothed, and highly polished rock surfaces seen in the higher parts of the Sierra. Overall, abrasion is less significant than the capacity moving ice has for plucking or pulling away (quarrying) big blocks of jointed bedrock. It's a lot easier for a glacier to pick up a block of rock weighing a ton and carry it away bodily than it is to laboriously wear it away by sandpapering.

Glacial excavation has created rock basins at the heads of Sierra canyons, now largely occupied by lakes or meadows. Rock basins were also formed farther down canyons where bedrock jointing was favorable for plucking. Most glaciated canyons have been widened by lateral plucking, giving them an open U-shape (see Photo 2–19) that contrasts with the V-Shape of stream-cut canyons. In places deepening and widening of main canyons carrying large glaciers has left tributary canyons "hanging." Today, streams descend from such hanging valleys in a succession of cascades and falls.

Although the floors of many glaciated valleys make pleasant hiking because of their gentle gradients, every now and then one encounters a steep precipitous bedrock rise as much as a thousand feet high. These are glacial steps. They usually occur where a fortuitous juxtaposition of well-jointed and relatively unjointed rock allowed the glacier to exercise great differential plucking.

Excavation at the heads of glaciers undercuts the slopes of high peaks and ridges, making them ragged, sharp, and narrow. Pointed peaks like the Matterhorn in the Alps have been given that shape by glacial excavation on three or more sides. Sharpening of divides has not happened everywhere in the Sierras, however, because in some instances ice flowed from the west side of the range eastward through passes across the range's crest. Such passes are broad, open, and smooth. Perhaps you have hiked through some and have wondered about them.

As the ice streams flowed down canyons they moved into a warmer environment where melting was more rapid, and eventually a balance was struck between the forward movement of the ice and rate of melting. Then the rock debris being carried by the ice was dumped in one place creating an *end moraine*. Melting also occurred along the lateral edges of the ice stream up the valley, and material dumped along the ice margin formed a *lateral moraine*. The Sierra Nevada have beautiful examples of end and lateral moraines, especially along the east base north from Bishop. Here the glaciers pushed out onto the gentle surface at the foot of the range, building huge embankments hundreds of feet high (see Photos P–6, 7, 8). These are located for you in the Mammoth trip guide.

BASIN RANGES

The Basin Ranges are rugged desert country with much topographic relief. The lowest point (Death Valley) is 282 feet below sea level, and the highest (White Mountain Peak) is 14,242 feet above. The Sierra Nevada, here treated as a separate province, are actually a basin range that contain Mt. Whitney (14,496), thus increasing the Basin Ranges relief by 256 feet. Whitney is only 82 airline miles west of Badwater in Death Valley. Even the local relief exceeds 11,000 feet, between Telescope Peak

(11,047) and the floor of Death Valley (–282). This means you can keep warm in winter by camping low, and cool in summer by camping high. Native Americans knew this long before Caucasians invaded the area.

The name "Basin Ranges" came about in a curious way. Much of far western United States is in the Basin and Range province, so named for its long, narrow mountain ranges separated by intervening valleys or basins. A subprovince of this region lying between the Sierra Nevada and the Wasatch Mountains, behind Salt Lake City, is designated the Great Basin because it does not drain to the sea. The area of concern here is part of the Great Basin. It is also part of the larger Basin and Range province with long narrow mountain ranges. Because these mountains lie within the Great Basin, they are designated "Basin Ranges."

Character and Dimensions

The triangular area of Basin Ranges in California is bounded on the west by the Sierra Nevada and on the east by the Nevada border. It is cut off abruptly on the south by the Garlock fault (Photo H–1). This triangle has an east-west base of 135 miles and a north-south dimension of 170 miles (Figure 2–13).

The mountain ranges and intervening valleys are 50 to 100 miles long by 15 to 20 miles wide, and they trend a few degrees west of north. Roughly half the terrain consists of mountains, which for the most part are dissected and rugged, although some—such as the Panamints, Cosos, and Inyos—have areas of gentle upland. For the most part, the ranges rise abruptly above the floors of intervening valleys with steep imposing faces. The 9,000-feet scarp of the Inyos west of Saline Valley is spectacularly steep and imposing. Principal among the ranges are the Inyo, White, Argus, Slate, Panamint, Cottonwood, Last Chance, Grapevine, Funeral, Black, and Amargosa mountains. Major valleys are Owens, Panamint, Saline, Eureka, Death, and Pahrump. Saline Valley has a breathtaking topographic closure of 3,000 feet. A lake would have to be that deep to overflow out of the basin. A similar lake would have to be more than 2,000 feet deep to overflow from Death Valley.

Most valleys are deeply filled with alluvium, and their fans rise gradually and gracefully from central dry-lake flats (playas) to an abrupt contact with the mountains. The ascent of these alluvial slopes gets surprisingly steep toward their heads; so don't assume something has gone wrong with your car if it seems sluggish on the final pitch.

Some valleys have clusters of low hills produced by recent deformation or scattered, isolated knobs representing older topography only partly buried by alluvium. A few have impressive sand dunes, such as those at the south end of Eureka Valley, 700 feet high. Dunes at the north end of Panamint Valley, in Death Valley, and elsewhere are smaller.

The only perennial flowing streams of any size are the Owens and Amargosa rivers. Both end up in local sumps, Owens River (before diversion into the Los Angeles aqueduct in 1913) in Owens Lake and Amargosa River in Death Valley.

Rocks

Like Mojave Desert, the Basin Ranges display rocks of many ages and types. There are Precambrian metamorphic rocks—mostly gneiss and schist, at least 1,800 m.y. old—and Late Precambrian rocks—up to 1,200 m.y. old—are represented by several thousand feet of weakly metamorphosed sedimentary beds, largely sandstone, shale, conglomerate, and carbonate layers. These rocks are locally intruded by basic igneous rock to produce talc deposits widely mined in the Death Valley region. Paleozoic rocks are abundant and widely distributed. They consist of well-layered sequences of sandstone, shale, and carbonate beds, aggregating tens of thousands of feet in thickness,

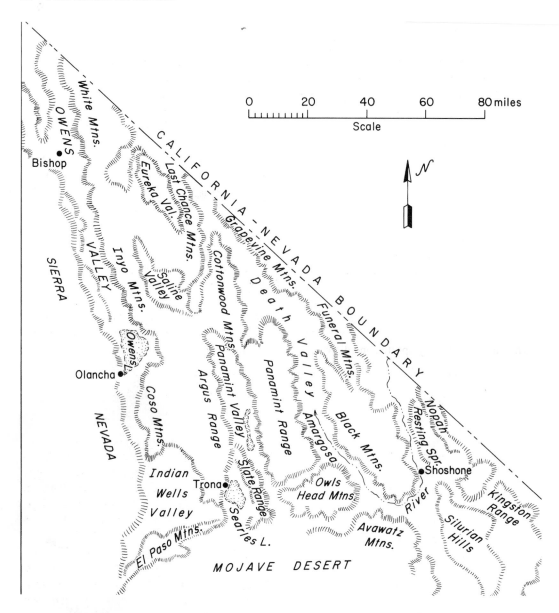

Figure 2–13.
Basin Range province.

locally fossiliferous. A considerable thickness of sedimentary rocks containing only scant microfossils or imprints of soft-bodied organisms lies between undoubted Precambrian rocks and proven Paleozoic beds in some California Basin Ranges. Mesozoic rocks are largely granitic intrusive bodies similar to the igneous rocks of the Sierra Nevada. Small areas of metamorphic rocks of Mesozoic age are known in the south, plus some small patches of sedimentary rock.

Cenozoic rocks occur as land-laid accumulations of sand, gravel, silt, and fragmental volcanic debris within the valleys and locally in mountains where they have been raised during mountain building. Cenozoic lava flows and associated beds of fragmental volcanic debris are common. Many alluvium-filled valleys are tilted by recent deformation, so that even very young alluvium has been locally uplifted and dissected.

Structure

By now you realize that most of the larger topographic features of our western lands are determined by faults and folds. The Basin Ranges are no exception. Each range is a fault block, bounded on one side, or both, by faults. The ranges have either been uplifted or the intervening valleys dropped. These movements are geologically young, having occurred mostly in the last few million years, and in places they still go on. Recent movement is indicated by fresh fault scarplets cutting alluvial fans (see Photo E–9).

Rocks making up the mountain ranges have a complex structure produced by episodes of deformation that occurred long before uplift of the ranges. The older rocks buried beneath alluvium

in the valleys presumably have sim complexities. Among these older huge low-angle faults, one of whic gosa—has had a hand in creati chaos within the rocks involved, in the southern part of Death Valley. (This Amargosa chaos is treated in the Death Valley trip guide.)

Extending northwesterly for at least 200 miles through the Death Valley region is a major fault complex, the Death Valley-Furnace Creek fault zone (see Figure 1–1). It roughly parallels the San Andreas and displays other similarities, such as right-lateral displacement and recent activity. Lateral displacement of several miles seems agreed on, but geologists disagree about the 50 miles proposed by some investigators. Regardless, this is a major structural feature, and it plays a large role in determining major topographic configurations within the region.

Resources

Historically this province has been southern California's major producer of silver, lead, and zinc, primarily because the abundant limestone and dolomite rocks of the area, where intruded by igneous bodies, provide favorable geological conditions for the deposition of minerals containing those metals. Cerro Gordo, high in the southern Inyo Mountains east of Keeler, Santa Rosa Mine, 10 miles south-southeast, and properties in the Darwin district between the Coso and Argus ranges have been the main sources. In early days the Panamint Range also supplied its share of lead and silver. Tungsten has come from Darwin and the Panamints, and modest amounts of copper, molybdenum, and gold were obtained as byproducts of ore refining for other metals. A little mercury was at one time recovered at Coso Hot Springs.

Among nonmetallic deposits, salines from the dry lakes and from Tertiary sedimentary deposits lead the way. Searles Lake is world famous for the richness and variety of its chemicals, principally salts of potash, boron, and lithium. Some bromine is obtained as well as common salt and carbonates and sulfates of sodium. Initially California's borax was produced mostly in the Death Valley area, and sizable deposits there have again been mined since 1988. Talc is obtained from mines scattered through the Inyos, Panamints, and the Death Valley area where it has formed in carbonate rocks close to basic igneous intrusive bodies. High-grade refractory material used in spark plugs was formerly mined in the White Mountains, and the Last Chance Range once sported a sulphur mine. Mineralization is widespread in this province, and it has been extensively prospected, but truly large metal mining operations have not been developed, except for Cerro Gordo.

Special Features

The Basin Ranges are rich in special features, such as salt pans, sliding rocks on playas (Racetrack), huge alluvial fans, remarkable turtleback structures (Death Valley), volcanic explosion craters (Ubehebe), and huge sand dunes (Eureka Valley), among others. Resolutely, we confine attention to an integrated chain of lakes (Figure 2–14) that once occupied parts of some of the principal valleys, 10,000 to at least 100,000 years ago.

During the great ice age, the Basin Ranges area was cooler and better watered. The mountains, which even today occasionally sport good blankets of snow, were then more heavily covered, and springtime thaws delivered much water to adjoining basins that sustained shallow lakes. However, do not forget our friend, the Sierra Nevada, to the west. This high massive range captured huge quantities of snow during the ice age, and great ice streams tens of miles long choked its valleys. Melting of this ice sent great quantities of water to both east and west. The Owens River on the east must have more than doubled its discharge. It filled the basin of Owens Lake at the south end of Owens Valley to overflowing. At that level, the expanded lake was over 200 feet deep (see Photo M–10) and covered 240 square miles. Its outflow stream ran south through Haiwee Meadows (now Haiwee Reservoir) into Rose Valley and cut the narrow defile at Little Lake forming a narrow gorge with rapids and a spectacular waterfall (see field guide Segment M).

Then it emptied into Indian Wells Valley now occupied by Inyokern, China Lake, and Ridgecrest. Here a broad shallow lake only 30 feet deep formed before overflow occurred eastward by way of Salt Wells Valley to Searles Basin. The vertical drop from Owens Lake to Searles Basin is over 1,500 feet. Waters became deeply ponded in Searles Basin, eventually attaining a depth of 640 feet and backing up Salt Wells Valley into Indian Wells Valley to make a single large water body covering some 385 square miles. This was the largest lake in the chain.

At the 640 foot depth, Searles Lake overflowed a divide at the southeast corner into Leach Trough. From here the water flowed eastward down the trough and then made a sharp turn north into the southern end of Panamint Valley. The lake formed in that basin was 60 miles long but only 6 to 10 miles wide. It covered about 275 square miles and had a depth approaching 1,000 feet.

At this level the water possibly overflowed by way of Wingate Pass (Photo 2–21) in the southern Panamint Mountains into Death Valley where it joined the Amargosa and Mojave rivers to make a body over 600 feet deep, named Lake Manly. More attention will be given to Lake Manly in the trip guide to Death Valley.

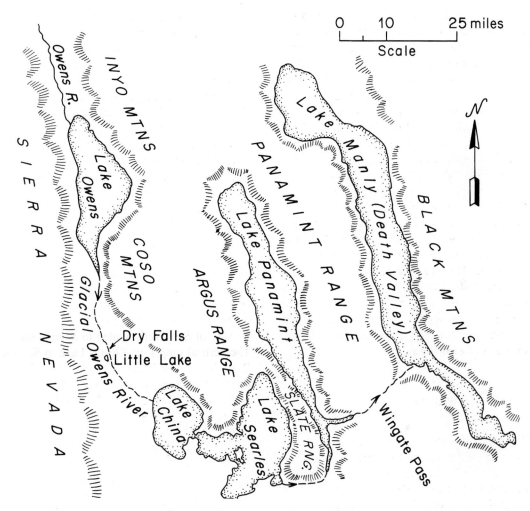

Figure 2–14.
Pluvial lakes fed by glacial Owens River runoff.

The shoreline features—cliffs and beaches—of these ancient lakes can be seen today in places around the margins of some of these basins (see Photo E–4). This country must have been extremely attractive when it was occupied by lakes and early Native Americans settled on lake-front sites.

We still get benefits from these lakes, especially in Searles Basin. They made up a system of huge decanting vessels and evaporation basins. The waters entering Searles Basin had attained a composition such that many chemically valuable salts were precipitated there when the lake dried up. This is the basis of the large chemical extractive industry operating today at and around Trona. The hundreds of large calcareous tufa cones and spires, to 140 feet high, at the south end of Searles Lake compose one of the weirdest landscapes in all the United States. They are well worth a visit. The road to them has recently been improved.

Photo 2–21.
High-altitude oblique view over Death Valley, looking south. (U.S. Air Force photo taken for U.S. Geological Survey, 374R-192.)

Trip Guides

Reading about geology is fine, but seeing some is better, much better. This chapter deals with geological relationships along selected southern California highways. Limitations of space require that choices be made. Two routes richly endowed with striking geological features and likely to be traveled by out-of-door people were selected; specifically Los Angeles Basin to Death Valley and Los Angeles Basin to Mammoth (Figure 3–1). The difficulties of driving in heavily populated areas dictate that the guides for these routes start from the periphery of the Los Angeles Basin.

Six supplementary spurs have been appended covering Baker to Las Vegas (including the Red Rock Recreation Area), San Bernardino to Palm Springs, and San Fernando Pass to San Joaquin Valley. All routes are divided into segments, listed in the table of contents and here, as well as being shown in Figure 3–1. Treatment of highways of the coastal zone, from the Mexican border north to San Luis Obispo, is given in Kendall/Hunt's companion volume *Coastal Southern California*.

USING THE TRIP GUIDES

The location of trip segments is shown on Figure 3–1. One-day excursions out of the Los Angeles Basin can be made by combining segments. For example a nice outing over and around San Gabriel Mountains can be made by putting Segments I and J together. Combining Segments A, G, H, L, K, J, or I would make a full day's adventure into Mojave Desert.

The following lists this book's route guide segments:

Los Angeles Basin to Death Valley	
A.	Devore to Barstow
B.	Barstow to Baker
C.	Baker to Shoshone
D.	Shoshone to Ashford Mill
E.	Ashford Mill to Furnace Creek Ranch

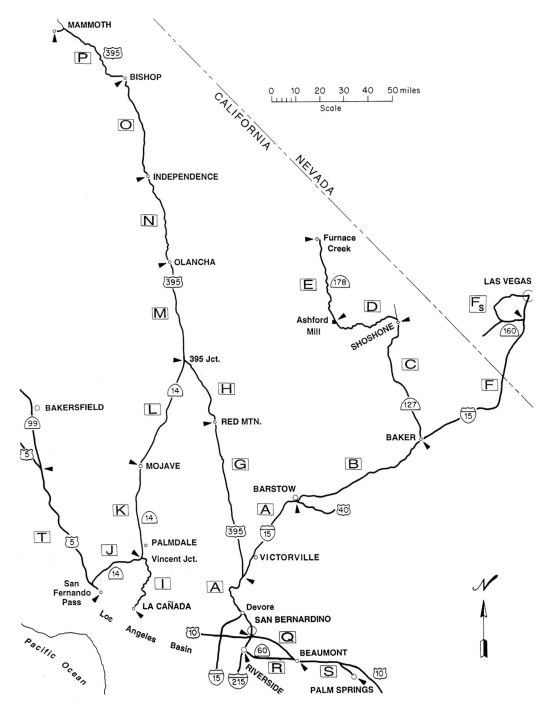

Figure 3–1.
Index map of field-guide segments.

Los Angeles Basin to Mammoth	
G.	Via Highway 395 to Red Mountain
H.	Red Mountain to Junction with Highway 14
I.	La Cañada to Vincent Junction
J.	San Fernando Pass to Vincent Junction
K.	Vincent Junction to Mojave
L.	Mojave to Highway 395
M.	Highway 395 to Olancha
N.	Olancha to Independence
O.	Independence to Bishop
P.	Bishop to Mammoth
Supplementary	
F.	Baker to Las Vegas (and F$_S$—Red Rock Recreational Areas [Nevada])
Q.	San Bernardino to Beaumont
R.	Riverside to Beaumont
S.	Beaumont to Palm Springs
T.	San Fernando Pass to San Joaquin Valley

Following a road guide is something of an art, but everyone improves with practice. It is helpful to read an entire segment before driving it. This alerts you to features ahead and provides a sense of succession and scale. If you have a travel companion, have that person read aloud as you motor along, and don't hurry. It is usually not possible to see everything the first time, so trips can be repeated with profit. Furthermore, conditions of visibility and lighting may make something you couldn't see the first time evident on a later occasion.

Because the Los Angeles Basin is the most heavily populated area, guides are written for travelers outbound from there. Inbound people have to work backward, page by page, through the text of a route segment, which is not always easy. Italicized inserts dispersed through the route descriptions provide help by identifying navigational points for inbound travelers.

The Art of Navigation

Navigators (land, sea, or air) need the best map they can get. The county road maps issued by automobile associations are good and should be obtained. Each segment herein is accompanied by a page-size, line-drawn map showing critical points and other useful information. Make friends with maps. Children rapidly become adept at navigation, their eyesight is keen, their memories good, and they enjoy the procedure; include them in the game by assigning a specific task.

A small number in a circle ③ on these maps identifies a special feature or a dissertation.

Professional road guides traditionally use odometer readings for location and navigation. This is an orderly and reasonably precise system, but it easily gets out of whack. Odometer readings are used more to give a general sense of scale. We would rather have your eyes focused on the passing landscape than glued to the odometer. We use tenths of a mile over short distances to keep you alert. At 60 miles an hour it takes just 18 seconds to cover 0.3 mile. Solid black triangles (▶) on page margins indicate desirable odometer check points. Black dots (●) indicate the start of a new subject or section along the route.

A useful means of locating features with respect to the direction of travel is to imagine a large clock laid flat on the road ahead with 12 o'clock directly up the road. Then 9 o'clock is directly (90°) to the left and 3 o'clock is directly to the right. This enables you to say, "Look at that big bull moose over there about 100 yards away at 11:30."

A host of roadside features and landmarks, natural or artificial, are used as navigational aids, but they should be reasonably permanent. Examples are buildings, power line or railroad crossings, road intersections, emergency call boxes (Figure 3–2), highway changes (divided versus undivided), pipelines, artificial dinosaurs

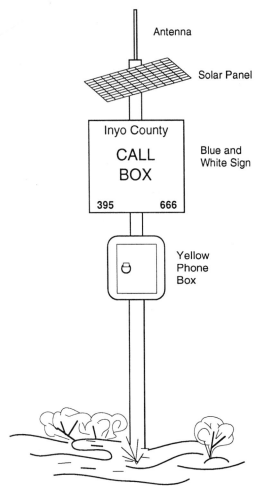

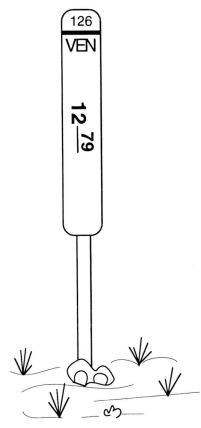

Figure 3–3.
Sketch of roadside mileage marker (MM). Mileage is vertical figure.

Figure 3–2.
Sketch of emergency call box. Critical number is *666* in lower right corner.

(no joke), and especially official highway signs, which are numerous and long lived. Highway signs that make a statement, such as *"Watch for Cattle Ahead"* or *"Weedpatch 7, Mud Pit 15, Broken Wheel 35,"* are identified in the guide text by quotation marks and italics.

Many secondary and some primary California roads have roadside mileage markers or mileage posts (called MMs or MPs), which are erected by counties. They are usually a white metal strip about 8 inches wide by 24 inches long mounted vertically on an iron stake pro-

jecting three or four feet above ground. Some bear clear or orange reflectors. The one illustrated (Figure 3–3) says this is Highway 126 in Ventura County 12.79 miles from some reference point. Be warned that mileage figures change abruptly as you cross a county line or change a highway. Markers are used occasionally to establish a precise location. Some bridge abutments and overpasses, along secondary roads mostly, bear mileage figures. See whether you can spot some. Practice navigation; it pays. We will give you all the help we can. Any road guide in California is likely to be out of date in places before it can be published. The geology does not usually change that fast, but highways and human artifacts do.

LOS ANGELES BASIN TO DEATH VALLEY

This guide begins at the intersection of interstate highways 15 and 215 near Devore. It proceeds to Furnace Creek Ranch in Death Valley via Barstow, Baker, Shoshone, over Salsberry and Jubilee passes, and up the Death Valley floor through Badwater to Furnace Creek Ranch, a total distance of roughly 275 miles (see Figure 3–1). A spur segment Ⓕ runs from Baker to Las Vegas, Nevada. It is best to have read the first two paragraphs of Segment A before you get to Devore, because things happen rapidly at the start.

Segment A—Devore to Barstow, 60 Miles

▶ | Note odometer reading on leaving the I-5/I-215 junction at Devore.

Beyond Devore our route parallels Cajon Creek on the southwest and converges slowly with the San Andreas fault to the northeast (Figure 3–4). All road cuts here would show good exposures of alluvial gravel and highly deformed metamorphic and igneous rocks, if they were not so overgrown by brush and grass. Watch for the Kenwood Avenue exit about 0.4 mile from I-5/I-215 junction. In another 0.5 mile you pass a large right-side road cut in dark slabby rock, the Pelona schist, which we will talk about shortly. The San Jacinto fault follows a nearly parallel course near Lytle Creek west of Cajon Creek. You are traversing the severely deformed block between it and the San Andreas, two of California's largest and most active faults.

Inbound: "Kenwood Ave. Right Lane" is near the Pelona Schist cut.

① About 2.5 miles from the I-5/I-215 junction the highway begins to curve right and soon passes through two huge, double-walled road cuts exposing mostly dark metamorphic rocks. Suddenly it bursts into more open terrain, the Cajon amphitheater. You have intersected the San Andreas fault zone. Look immediately northwest at 9:30–10 o'clock (not the driver) to get a striking view directly up Lone Pine Canyon (Photo A–1), a linear cleft defining the trace of the San Andreas fault.

Inbound: The San Andreas is crossed 2.4 miles beyond Cleghorn Road.

● Within a mile the railroad cuts low down, across Cajon Creek at 9 o'clock, expose light-colored crystalline rocks in contact at the north end with a fault slice of well-layered, dark-brown marine shales of the San Francisquito Formation (60 m.y.). Beyond, in cuts and on the hillsides at 9–11 o'clock are extensive exposures of the pinkish Cajon Valley beds displaying folded and faulted strata inclined in different directions (Photo A–2). These middle to upper Miocene (6 to 15 m.y.) beds consist of moderately indurated, nonmarine deposits of sandstone, conglomerate, and shale, totaling up to 9,000 feet thick. Formerly they were correlated with similar deposits on the opposite side of the San Andreas fault in Devils Punch Bowl near Valyermo, 23 miles to the northwest. That interpretation is now questioned on the basis of sedimentological and age differences. If they were correlative, the two units should be much farther apart, judging from other indications of displacements along the San Andreas.

Inbound: Cajon Valley beds are well exposed on the right beyond Cleghorn Road. The San Francisquito exposure is at 3 o'clock 1.1 miles beyond Cleghorn Road.

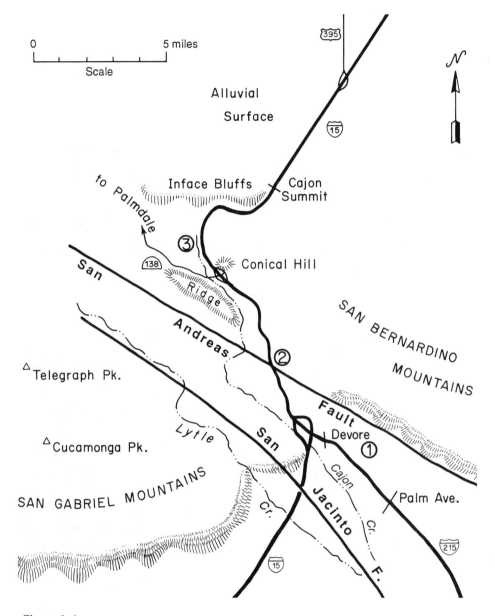

Figure 3–4.
Segment A, south half, Devore to Highway 395 separation. Circled numbers such as ①
are keyed to sections in text.

Photo A–1.
Looking west-northwest directly up Lone Pine Canyon marking the trace of San Andreas fault, as seen from Interstate 15. Highest snow capped Peak is Old Baldy.

Photo A–2.
Exposure of folded Cajon Valley beds northwest of Cajon Creek just across San Andreas fault as seen from Interstate 15. (Photo by Helen Z. Knudsen.)

● At and beyond the Cleghorn Road intersection, look across Cajon Creek at 9:30–10 o'clock to see a high, somber, gray crystalline-rock ridge behind lower exposures of pinkish Cajon Valley beds. This ridge is a fault wedge, or slice, forming an elongated island of crystalline rock surrounded by Cajon Valley beds. It parallels the freeway for the next two miles.

Inbound: This ridge is seen between 2 and 3 o'clock near Cleghorn Road. Note odometer reading at Cleghorn overpass.

▶ Note odometer reading after weigh station.

Photo A–3. Outbound view of conglomeratic sandstone hogback or flatiron ridges (Mormon Rocks) in Cajon Valley beds just beyond Cajon Junction on I-15. (Photo by Helen Z. Knudsen.)

② Passing the Weigh Station, look ahead low down at 11–11:30 o'clock to see large, light-colored, hogback and flatiron ridges, locally known as Mormon rocks (Photo A–3), formed by erosion of well-cemented, tilted Cajon Valley beds. Beyond the Weigh Station, approaching Palmdale turnoff at Cajon Junction, the conical hill (see Figure 3–4) lying just east of the freeway at 1 o'clock is either a plate of crystalline rock thrust out of the San Andreas fault zone onto the Cajon Valley beds or an island of crystalline rock projecting through Cajon Valley beds that have been deposited around it.

Inbound: The flatiron ridges are seen at 3 o'clock approaching Cajon Junction, the conical hill is at 11:30 at "Palmdale Hwy. 138, exit 1 mile."

● About two miles beyond the Palmdale exit you start to climb out of the Cajon amphitheater, an open basin created through erosion by Cajon Creek and its tributaries. It exists because the sedimentary rocks on the northeast side of the San Andreas fault are soft and susceptible to rapid erosion, compared to the hard crystalline rocks southwest of the fault. Once Cajon Creek

worked its way across the fault, it had a field day removing these softer materials, principally the Cajon Valley beds and a succession of weakly consolidated overlying sedimentary units.

Inbound: Views of Cajon amphitheater are great descending from Cajon Summit.

Ascending the grade you look north against the high Inface bluffs carved by the headward growth of Cajon Creek eating north into the desert country. Shortly the freeway swings east and obliquely traverses this face. The upper-most part of the bluffs consist of fan gravels carried north at earlier periods from the San Gabriel and San Bernardino mountains. These gravels contain fragments of Pelona schist that could have come only from south of the San Andreas fault. They had to be carried here by streams before Cajon amphitheater existed. Un-fortunately, weathering has so homogenized the large road cuts ahead that you can no longer observe the striking nature of these deposits.

Inbound: Big road cuts in gravels of Inface bluffs extend from Summit (4,260 feet) to "4,000 feet elevation."

● You pop over Cajon Summit at 4,190 feet, according to the sign, a little over 4 miles beyond Palmdale exit. Vegetation at the summit is a scraggly chaparral of scrub oak and chamise. Our freeway and three railroad lines use Cajon Pass to get out of the greater Los Angeles area. This route was made possible by the erosional work of Cajon Creek taking advantage of the rock and structural patterns provided by the San Andreas fault. Without this unusual arrangement the passage from San Bernardino to the desert would be steep and difficult, probably impossible for rail lines.

Inbound: Views of these relationships are better inbound.

● Once over Cajon Summit and 0.3 mile past the Oakhill Road overpass, you start to descend a long, gently inclined alluvial slope toward Victorville, here dissected because of continuing slow tectonic uplift of the Cajon Pass area. This is part of the apron of fans that once extended south to the San Gabriel and San Bernardino mountains before being eaten away and beheaded by Cajon Creek. That was the time at which the Pelona schist stones were delivered into the fan gravels. At *"Victorville 14, Barstow 45,"* note the vegetation of scrub oak and joshua trees. A half mile farther there are more joshua trees plus junipers.

● About three miles beyond Cajon Summit, U.S. 395 separates from Interstate 15. Guide Segments G and H describe scenes along 395 to a junction with Highway 14, which is followed by the Los Angeles to Mammoth route (Figure 3–1). Travelers going to Death Valley and Las Vegas continue on Interstate 15 (Segment A), and those going to Mammoth turn onto U.S. 395 and skip to Segment G.

Inbound: Between 395 separation and Cajon summit note vegetation changes and also the increase in dissection.

● From the Highway 395 separation, continue north-northeasterly on Interstate 15 toward Victorville (Figure 3–5). Vegetation becomes more open and features joshua trees and junipers. Mojave River, draining from high, relatively well-watered parts of San Bernardino Mountains, flows north in a wide sandy bed about 10 miles to the east. You are slowly converging to a rendezvous with it at Victorville. Note that joshua trees become predominate beyond *"Hesperia 1 mile."*

Inbound: Joshua trees and creosote dominate before Hesperia exit, but after the exit it is joshua and junipers.

● Beyond the Hesperia-Phelan exit, the west end of the broad Lucerne Valley trough extending from Victorville east-southeasterly to the Twentynine Palms country makes the low skyline at 2 o'clock (see Mojave Desert province). The east branch of the California Aqueduct is crossed in 0.4 mile. Ahead are the bedrock hills of Victorville, composed of granitic and metamorphic rocks. Some of the white spots are carbonate-rock quarries (limestone and marble) providing material for the cement plants of Victorville and Oro Grande. At 10 o'clock are low peaks and ridges of the Shadow Mountains. They contain extensive exposures of metamorphic and igneous rocks like those near Victorville. At *"Victorville Next 5 Exits"* the west end of Lucerne Valley trough is prominent at 3 o'clock.

Inbound: These features are seen by looking to the sides and back between Victorville and the Hesperia exit.

● Beyond the Lucerne Valley exit, plumes of dust and smoke from cement plants at Victorville and Oro Grande are sometimes visible. Smoke frequently seen dead ahead beyond Victorville is from the city dump. At 11 o'clock are granitic rocks at the base of Quartzite Mountain, and the conical hills at 1 o'clock are erosion residuals carved mostly in metamorphic rocks.

Inbound: You have already seen this area before crossing Mojave River.

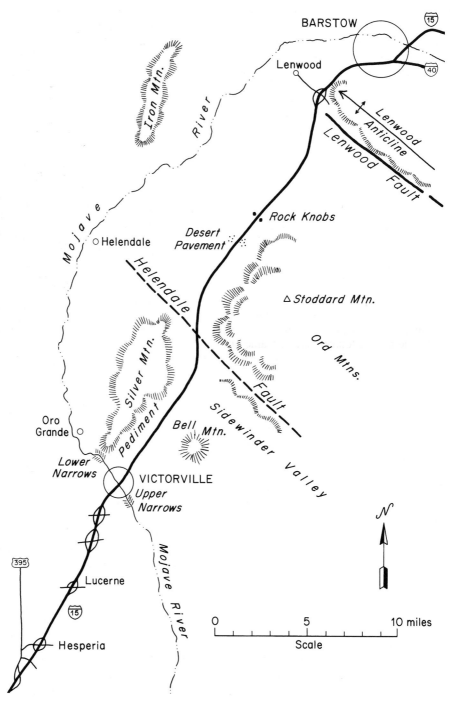

Figure 3–5.
Segment A, north half, Highway 395 separation to Barstow.

● As the freeway curves around Victorville you become aware of greater gullying and dissection. This is caused by the Mojave River, which flows in an entrenched course 175 feet below the alluvial surface on which you have traveled.

Inbound: To be seen once the highway rises to the smooth alluvial surface on which roadside buildings sit.

● Shortly you drop down and cross Mojave River.

▶ | Note odometer reading at crossing.

A good stream of water flows here all year long, although the river bed a few miles upstream is bone dry except during floods. What happens is this. About 1.5 miles upstream at the Apple Valley bridge (Upper Narrows, see Figure 3–5), where the Mojave River became entrenched, it uncovered a westward projecting ridge of granitic rock buried in the alluvium. The river had no choice but to carve a narrow, steep-walled gorge into the granite, creating the Upper Narrows (Photo A–4). Bedrock under the river bed at this point brings subsurface water from upstream that is seeping through the porous river bed sand back to the surface. The water disappears again a mile or so below the Lower Narrows (see Figure 3–5), which were formed in the same fashion.

Inbound: You are over the water channel of Mojave River at 0.8 mile beyond Stoddard Wells Road exit.

● Across the river you start climbing and good exposures of jointed granitic rock are soon seen at 1–5 o'clock. They are discolored gray by a mantle of cement dust that has been converted to a hard surface coating. To the left at 9–10 o'clock you see again the granitic rock flanking Quartzite Mountain. The gently sloping bedrock surface extending outward (south and southeast) from the mountain base across granitic rock has been formed by erosion and is

Photo A–4.
Upper Narrows of Mojave River at Victorville, as viewed southeastward. Formed as the river cut down into a buried ridge of granitic rock, now exhumed by further erosion. (Photo by John S. Shelton, 3402.)

called a *pediment*. Within two miles beyond the river, quarries, roads, and other workings in marble (light colored) are visible among hills to the right. At about four miles you begin to traverse the broad alluvial surface of well named Sidewinder Valley. It harbors a good number of little, ill-tempered rattlesnakes.

Inbound: Discolored granite outcrops are seen at 8–10 o'clock between "Stoddard Wells Rd. exit 1 mile" and the actual exit. Quartzite Mountain pediment is at 3–4 o'clock.

● Beyond "*Speedometer Check*" all but the driver can look back at about 4:30 o'clock for a view of the graceful symmetry of Bell Mountain (Photo A–5) on the near skyline. The smooth concavity of its slopes reflects the fact that its summit is capped by metavolcanic rocks

Photo A–5.
Smooth, concave slopes of Bell Mountain in Sidewinder Valley are striking. (Photo by Helen Z. Knudsen.)

that yield particles of all sizes when the rock is broken up by weathering. These particles are distributed down the flanks of Bell Mountain in such a manner (coarser near the top and finer near the bottom) as to produce a slope grading smoothly from steep to gentle, hence concave. In another mile or so the rugged Stoddard Mountain mass (4,894 feet) in the Ord Mountains begins to dominate the scene at about 12 o'clock. The rocks therein display considerable variation in color and appearance being a mixture of intrusive granitic and older metavolcanics of possible Triassic age (230 m.y.).

Inbound: Bell Mountain is seen well at 10 o'-clock from roadside mileage marker (MM) 50.00. This is 1.8 miles beyond Boulder Road overpass.

● Shortly the highway curves west, passes under some large power lines, and in 1.8 miles passes through a double-walled road cut in hard, fine grained, fractured metavolcanic rock. In another 0.2 mile it starts a gentle descent into a shallow valley. You are approaching a crossing of the Helendale fault, one of a dozen major northwest trending faults slicing this part of the Mojave Desert like a loaf of bread (see Mojave Desert province). Another fault in this group, the Camp Rock about 20 miles to the east, was active at the time of the notable 7.4

magnitude Landers earthquake in June 1992. You are here traveling over metavolcanic rocks and debris derived therefrom, and a suspicion that a fault exists here arises from the abrupt change to light colored knobby granitic rock seen a little before crossing the bridge over Wild Wash, about 3.2 miles from the power lines. A few miles to the southeast, and also in Lucerne Valley, Helendale fault is marked by scarplets breaking alluvial fans, an indication of fairly recent movements. By looking back to the southwest at *"Wild Wash Road Exit, 1 Mile,"* you can see a similar low scarp here, better viewed inbound.

Inbound: Scarp of Helendale fault is seen at 11–2 o'clock 1 mile beyond Wild Wash overpass.

● For several miles beyond the Wild Wash bridge, note the blocky, angular character of rock outcroppings and of rock debris on the hill slopes east of the freeway. This is characteristic of the metavolcanic rocks of the Triassic Sidewinder Series widely exposed in these hills. Locally there are small bodies of intrusive granitic rocks that you may be able to recognize by the more rounded nature of their outcroppings.

Inbound: You see these rocks between Hodge Road and Wild Wash bridge.

Photo A–6.
Bedrock knobs of weathered granitic rock that look like piles of large boulders; alongside I-15 approaching Barstow. (Photo by Helen Z. Knudsen.)

● Approaching Hodge Road one gets good views to the left of the western Mojave, a region of low relief, broad domes, and small residual rocky peaks and ridges. The Mojave River is in a course about eight miles west, and you intersect it again at Barstow. The black rocks seen far off at about 10–10:30 o'clock are part of Iron Mountain. They are dark igneous intrusive rocks relatively rich in iron and magnesium.

> ▶ Note odometer reading at Hodge Road overpass.

Inbound: This scene is to the right beyond Hodge Road overpass.

● About 0.4 mile beyond Hodge Road overpass you come into areas where alluvial surfaces on both sides of the road have patches of desert pavement (dark, smooth, stony, vegetation-free areas). Stones within the pavement are blackened by rock (desert) varnish. Desert pavement is a concentration of tightly fitted stones left on the surface, mostly of alluvial deposits, as larger surface stones break up and as finer materials have been carried away. Slight dissection by gullies draining to the entrenched course of the Mojave River is slowly destroying this pavement, leaving only patches. The vegetation has become scattered creosote bushes.

Inbound: Pavement patches are a little beyond "Hodge Road 3/4 mile" and MM 61.00.

● A little more than 3 miles from Hodge Road you pass what looks like piles of huge rounded boulders on both sides of the highway (Photo A–6). These are actually the outcrops of jointed granitic bedrock rounded by weathering and granular disintegration, a typical behavior for uniform, coarse-grained igneous rocks in a desert environment.

Inbound: Rock knobs are 2.7 miles beyond Sidewinder Road, 0.5 mile beyond MM 64.00.

● At and beyond Sidewinder Road look dead ahead about three miles to a low, gray, skyline ridge with many small gullies. The ridge is a geologically young anticline (an upfold) within relatively unconsolidated alluvial-fan materials. This is the Lenwood anticline, and you will cross its western plunging nose as the freeway curves into Barstow beyond the Lenwood exit. To complicate matters, another northwest trending fault, the Lenwood, similar to the Helendale, passes along the southwest side of the anticline (see Figure 3–5).

Inbound: The scene is behind you. Pull off at Sidewinder Road for a look at it.

● Soon the highway bends westward, and buildings at the Lenwood exit are ahead. The smooth subdued skyline at 12:30 o'clock is near Rainbow Basin, the site of the Barstow syncline (a downfold) formed in the famed Barstow Formation, a middle Miocene (13 to 18 m.y.) terrestrial basin deposit rich in fossil remains of extinct vertebrate animals (see Mojave Desert province, special features).

Inbound: Unfortunately, you see mainly the manufacturers' outlets and huge truck servicing facilities here.

● About 1.5 miles beyond Lenwood exit, the freeway crosses the nose of the Lenwood anticline, and road cuts and hillsides expose some of the gently dipping beds of largely unconsolidated materials that compose this structure.

Inbound: At "Victorville 29, San Bernardino 69, Los Angeles 113," you are circling the nose of the Lenwood anticline. Good exposures of tilted beds are just beyond "Lenwood exit 1 mile."

● Approaching and beyond the West Main Street exit to Barstow, at 10–11 o'clock, is a prominent reddish-brown rock knob with microwave relay towers and a white letter B. This knob is a small intrusive plug of Tertiary igneous rock (rhyolite) formed at a time when the surface volcanics of the Barstow region were being extruded. It has subsequently been etched out by erosion.

Inbound: Barely seen just beyond the Barstow Road, Highway 247 exit.

● Beyond the Barstow Road exit and the I-40 off ramp to Needles you are descending toward a crossing of Mojave River.

Segment B—Barstow to Baker, 61 Miles

● Just beyond East Main Street exit at Barstow, the broad, sandy, normally dry bed of the Mojave River is crossed (Figure 3–6). At times of high flooding, a good stream of water flows here.

Inbound: Nobody can miss the bridge.

● At the Bakersfield exit (Highway 58), another northwest trending fault crosses your route but without recognizable expression at the freeway. Low hills traversed for the next three miles, consist of a mixed and strongly de-

formed assemblage of Tertiary volcanics and locally well-bedded, terrestrial sedimentary rocks. Road cuts in this reach show considerable deposits of *caliche,* a surficial white calcium carbonate deposit of secondary origin.

Inbound: These features are seen between the Fort Irwin Road and Highway 58 exits.

● Descending to the Yermo plain beyond Meridian Road, the clearly labelled Calico Mountains are in view at 9–11 o'clock. The name comes from their variegated appearance produced by complex structural intermixing among highly colored sedimentary and volcanic rocks of Tertiary age. The cream and green colors are sedimentary and volcano-sedimentary units, and the dark reddish-browns are principally volcanics.

Inbound: Look to the right approaching Ghost Town Road.

● Approaching Ghost Town exit, the large black knob just off the highway at 3 o'clock is Elephant Mountain. This is another near-surface intrusive plug, in this instance of a darker rock than the reddish plug in Barstow. It doesn't look much like an elephant from here, but from the east on I-40 between Newberry Springs and Daggett, the resemblance to a prone elephant, trunk and all, is strong.

▶ Note odometer at the Ghost Town Road underpass.

Inbound: The prone profile of the elephant was seen from near the inspection station.

● Beyond Ghost Town Road where the freeway rises over an underpass, look left to the mountains at 9:15 o'clock up the canyon behind the green trees and under the white "Calico." There you see buildings of the old mining town of Calico, now operated for tourists by San Bernardino County. Calico was a silver camp from 1882 to 1896, with a reported production of 86 million dollars, and then a borax producer until about 1907. The silver deposits

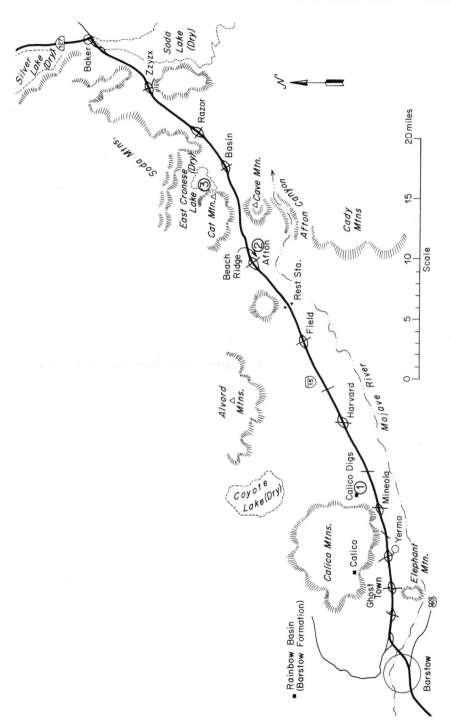

Figure 3–6.
Segment B, Barstow to Baker.

were rich, but they pinched out at shallow depths. Roads crisscrossing the mountain face west of Calico were made for core-drilling explorations by a large mining company searching for silver ore. The project is on hold because of environmental considerations. South of Calico Mountains is another member of those northwest trending fractures, the Calico fault.

Inbound: Look right to 3:15–3:30 o'clock at and beyond "Ghost Town Road exit, 1 mile."

● At *"Yermo Exit 1 mile,"* note the highly colored beds in the canyon cutting the mountain front at 9 o'clock. In the opposite direction at 2:45 o'clock, the high tower with a white box on top is part of an experimental Southern California Edison Company solar collector arrangement for generating power. The white box is at the focus of a large number of reflecting mirrors on the ground.

Inbound: Seen within the first mile beyond the Yermo exit.

● Skip ahead to ① and read about the archaeological diggings. To visit the Early Man site, turn off on Mineola Road, about a mile beyond the agriculture inspection station, turn north across the freeway, drive 0.5 mile east on the paved road, go north 1.1 miles on the wide graded dirt road, and then turn east on the one-track labeled road. Visitors are welcome, and you should find the experience fascinating and educational.

Inbound: Turn off at Mineola Road to visit archaeological site.

● Beyond the agricultural inspection station you can look at 2 o'clock down the large trough extending from Barstow to beyond Amboy. This is the route followed by Interstate 40, the Santa Fe Railway to Needles, and possibly in earlier times by the Mojave River, which now has a course closely parallel to the one you travel.

Inbound: Look back beyond Mineola Road to see Amboy Trough. West of (beyond) the inspection station the prone profile of Elephant

Mountain at 11 o'clock is pretty good, and the Edison Company solar tower is also in view at about 10:30.

① Just before you pass under the large power lines ahead, look two miles north at 9 o'clock. The pale area on the hill-face with trailers and small buildings is the site of the "Calico Digs," an area excavated by the San Bernardino County Museum of Natural History. At this spot some possibly human-shaped primitive artifacts (stone choppers) have been recovered from deep pits dug in well-cemented fan gravels. If accepted as genuine, they establish occupation of North America by ancient humans much earlier than generally thought, at least 50,000 and perhaps many more years ago.

Inbound: "Calico Digs" are seen to the right just after passing under the four large Boulder Dam power lines.

● About 1.5 miles beyond the power lines, low hills close to the freeway on the north side expose inclined, cream and greenish mid-Miocene Barstow beds capped by younger alluvial gravels (Photo B–1). In another 1.5 miles, beyond the next overpass, the west end of the Cady Mountains is at 1–2 o'clock. They are here composed mostly of dark Miocene volcanic rocks, extensively mantled by light-colored, wind-blown sand and silt picked up from the Mojave River alluvial plain by the prevailing westerly wind. Little knobs on both sides of the freeway in the next several miles are composed largely of Barstow beds (light) and volcanic rocks (dark).

Inbound: At MM 93.00 Calico Hills come into view ahead, beyond an overpass with no exits are knobs of Barstow beds, and larger exposures of same are to the right approaching the power lines.

● Just before *"Harvard Road 1 Mile"* is a pistachio-nut orchard south of the freeway. This area and the broad flat of the Mojave Valley south of the railroad and freeway, with its alfalfa and hay fields, is well endowed with ground water thanks to infiltration from the Mojave River.

Photo B–1.
Exposure of gently dipping Barstow Formation beds just north of I-15 a little east of the crossing of four large power lines east of Mineola Road interchange. (Photo by Helen Z. Knudsen.)

▶ | Note odometer at the Harvard Road overpass.

Inbound: The pistachio-nut orchard is a mile west of the Harvard Road overpass. Lake Dolores and its water slides are on the right about 2.5 miles beyond Harvard Road overpass. Beyond the overpass with no off ramps, rock knobs immediately to the right are mostly tilted mid-Miocene (15 m.y.) Barstow beds.

● About 0.2 mile beyond the Harvard overpass is a double walled road cut in gravel with bedding tilted 15°. Then 1.5 to 2 miles farther, Alvord Mountains can be seen about 10 miles away at 9–10:30 o'clock. The light gray materials on the skyline at their far end are uplifted, gently deformed Quaternary fanglomerates (somewhat consolidated alluvial fan deposits). They rest on darker Tertiary volcanic rocks of complex structure, which in turn overlie older granitic rocks that have intruded a still older metamorphic sequence exposed mostly toward the near end of the mountains. Irregular light areas within the dark near end rocks are surficial deposits of wind-blown sand and silt.

Inbound: Beyond the swale, Alvord Mountains are good on right skyline. The double-walled road cut is near "Harvard Road exit 1 mile."

● Nearly 5 miles beyond Harvard Road, and 1.7 miles beyond a no-exit overpass, approaching a southward curve the freeway dips into a broad shallow swale. Here to the right are some dissected, soft silty beds deposited in a body of water formerly covering 200 to 300 square miles in this basin, named Lake Manix. Large bodies of water covered this area many times within the last 185,000 years when the San Bernardino Mountains had more snow and when the runoff to the Mojave River was greater. Wet intervals in desert regions are called *pluvial periods,* a handy term. Pluvial Lake Manix was a little more than 200 feet deep. Its basin was probably created by deformation, perhaps faulting, but it was eventually breached by overflow and downcutting along Afton Canyon (see Figure 3–6).

The silts here and farther along were laid down around 15,000 years ago.

Animal life was abundant along Lake Manix shores. There were shellfish, turtles, beetles, and larger animals including early dogs, bears, cats, mammoths, horses, camels, antelopes, bison, and sheep. Their fossil remains are found mostly in the shoreline deposits. Among the abundant birds were pelicans and flamingos. The image of a pink flamingo standing stiffly at attention on one leg in the Mojave Desert is a bit incongruous.

Inbound: Swale with lake beds is about two miles west (beyond) Field Road overpass. It has two bridges in the bottom.

● More than a mile beyond Field Road overpass the prominent mountain at 11:30 o'clock is a mixture of old granitic and metamorphic rocks. The pale rock is unvarnished marble; the dark rock is a heavily varnished metamorphic.

Inbound: This mountain is opposite and to the right from the rest stop.

● Beyond the rest area, the high sharp skyline peak at 1 o'clock is Cave Mountain (3,585). Extensive small-scale gullying of the hillslopes seen far across the valley at 1–2 o'clock from near the rest area suggests relatively unconsolidated deposits, presumably largely fanglomerates. About 1.5 miles beyond the rest area are greenish lake beds across the railroad. At 1.7 miles from the rest area, the commercial plants at 3 o'clock along the railroad process materials trucked out of neighboring desert areas, particularly talc from Death Valley. Only the eastern and larger plant appeared active in 1993.

Inbound: Plants are at 9 o'clock at "Rest Area 2 Miles."

● Shortly one looks down Afton Canyon at 2 o'clock. This canyon was cut by overtopping outflow from Lake Manix about 15,000 years ago and is now the course of the Mojave River as well as the Union Pacific Railroad. Bedrock in the channel floor forces water to the surface, and short reaches of flowing stream usually exist in the canyon throughout much of the year.

Inbound: You have to look back to see Afton Canyon from "Rest Area 2 Miles."

● In about a mile, starting 0.4 mile beyond "Afton Road Exit 1 mile," is another and more extensive area of dissected Lake Manix beds. These pale-green, fine silts are capped by younger, brownish alluvial gravels. You can get a closer look by driving back west on the frontage road and walking down a gully cut in the lake beds.

Inbound: This is easy for travelers who have crossed to the south side of I-5 at Afton Road.

② Approaching the overpass at Afton Road, the freeway rises and passes through a magnificent gravel beach ridge formed along one of the higher levels of Lake Manix (Photo B–2). This is a good place to turn off for a look around. The ridge is hard to see from the entrenched freeway route, but if you don't turn off, look back beyond the overpass at 3–4 o'clock to see the backside of the beach ridge and the little playa flat it encloses. If you turn off, these relations are more obvious. The road to Afton Canyon campground extends southeast along the crest of the beach ridge from the boulevard stop.

▶ | Note odometer reading here. |

Inbound: It is worthwhile to pull off at Afton Road, cross to the south side of I-5, and park on top of the beach ridge.

● Back on the freeway, starting a mile beyond Afton Road, blocky granitic rock with a modest coating of desert varnish is exposed in hill slopes on both sides for the next few miles. Much wind-blown sand and silt from the west have accumulated here.

Inbound: You are in the sand-mantled area at MM 115.50.

● About three miles from Afton Road overpass, the freeway starts a descent to Cronese Valley and East Cronese Dry Lake, between Cave Mountain on the right and Cat Mountain on the left. At "Baker 24, Las Vegas 116,"

Photo B–2.
Beach ridge of ancient Lake Manix near Afton Canyon (in background), two miles south of I-15, as viewed southward from the air. (From *Geology Illustrated* by John S. Shelton, W. H. Freeman Company © 1966.)

Cave Mountain is at 12:15 o'clock. It is composed of an old tough type of granitic rock, which is why it is high, rugged, and darkly varnished. Note the rough, bouldery fans along the base of Cave Mountain, built at least in part by debris flows. The light-colored areas on these fans have been bulldozed.

Inbound: Bouldery Cave Mountain fans start near MM 117.50 and extend nearly to MM 116.00.

● Nearing the floor of Cronese Valley about seven miles from Afton Road, note that rock exposures close to the road on the south are lighter than rocks up the slope. Their rock varnish has been removed by the blasting of wind-blown sand that now partly mantles these lower slopes. Trees bordering the freeway on the right here are palo verde and salt cedar.

Inbound: Seen in the first mile beyond bridge.

● At the concrete bridge 8.8 miles from Afton Road, don't be surprised to see a small flowing stream at times in winter. This is Mojave River water that comes to the surface in Afton Canyon or flood water from the San Bernardinos. Upon leaving Afton Canyon, the Mojave can flow east to Soda Lake or north to Cronese Lake. At times of flood it often does both. In 1916 Mojave River flood waters accumulated in East Cronese Lake to a depth of 10 feet. The abrupt mountain face east of Cronese Valley is the west face of Soda Mountains, an irregular mass extending to Baker. The part viewed here is mostly granitic rock with a mantle of lava on the south flank.

Inbound: This scene is in good view between "Basin Road 3/4 mile" and the bridge.

③ Between 1 and 1.5 miles beyond Basin Road passengers can look back at 7:30–8 o'clock to see a large dune on the east face of Cronese Mountain. It was built of sand blown over the top of the mountain from the west. The shape, as viewed in favorable light, strongly resembles a cat lying on its tummy with ears and tail in their proper places (Photo B–3), hence the names "Cat Dune" and Cat Mountain.

Inbound: Cat Dune view is good from the vicinity of MM 123.00.

● Rocks in the mountains around Rasor Road are mostly Tertiary volcanics, locally mantled by wind-blown material. Within the mile beyond Rasor Road is a good view at 3 o'clock of Kelso Dunes 25 miles to the southeast. These dunes, fully 500 feet high, lie at the southeast end of Devils Playground, a barren sandy windswept plain into which sand is driven for 20 miles from the mouth of the Afton Canyon by prevailing westerly winds. Sand also comes from the north off Soda Lake flats. The dunes have accumulated at a site where the local topography causes strong storm winds from the north, south, and east to counterbalance the prevailing westerly wind. These beautiful dunes are accessible by road from either Baker or I-40 north of Amboy.

Photo B–3.
Cat dune, named from the resemblance to a cat lying on its tummy, ears and tail in proper places and proportions. This is a falling dune, built of sand blown over the top of the Cronese Mountains and accumulated on their eastern slope. As viewed westward from I-15 ascending from Cronese Valley. (Photo by Helen Z. Knudsen.)

Inbound: Kelso Dunes are seen in the far distance at 9 o'clock from "Razor Road exit 1 Mile."

● Looking ahead for the next 2 to 3 miles, hills right of the freeway at 12–3 o'clock are part of the Soda Mountains, and the rocks are granitic. Left of the freeway, at 10:30–12 o'clock, are subdued, gray hills composed of uplifted young Avawatz fanglomerates within which are local patches of dark heterogeneous rocks. These patches are mostly slices of Jurassic-Triassic (150 to 230 m.y.) metamorphic rocks inserted into the younger fanglomerates by faulting within the wide, complex, north-trending Soda-Avawatz fault zone.

Inbound: These features are seen to the right within 1 to 2 miles beyond Zzyzx overpass.

Smoke trees grow in the wash close to the highway 0.75 mile west of Zzyzx Road.

● After passing Zzyzx Road, Kelso Dunes are seen again at 2 o'clock in the far distance, and the floor of Soda Lake appears in the midforeground (Photo B–4). Soda Lake is left over from a much larger pluvial water body, Lake Mojave, which lay in this basin 10,000 to 15,000 years ago. White parts of Soda Playa are usually dry. Brown parts are kept moist by capillarity.

Inbound: These views are seen to the left approaching Zzyzx Road.

● In less than a mile the freeway passes through deep cuts in Avawatz fanglomerates earlier mentioned. From the highway beyond one gets good views of Soda Lake and the Devils Playground farther south at 4 o'clock.

Inbound: The road cuts are 0.2 mile beyond "Zzyzx exit 1 mile."

● Coming down the long straightaway toward Baker, the even skyline at 1–1:30 o'clock is made by lava flows, and the dark conical peaks at 2 o'clock are a few of the 40 volcanic cones within an extensive volcanic field in that area (Photo B–5).

Inbound: You have to look back at about 7 o'clock to see these volcanic features.

● Take the first turnoff into Baker, and as the highway climbs toward the freeway overpass, look to the left side of the highway to see a pit in the bedded fine gravels of a Lake Mojave beach.

The dark rock knob just left of the road coming into town is Baker Hill. The upper half is composed of badly broken and deformed upper Paleozoic (Permian) limestone about 250 m.y. old. Note many small niches and caverns formed by weathering in that rock. The lower half is without cavernous weathering and has a somewhat variegated lighter color. It is also much older, Precambrian in age. A low-angle thrust fault separates the two rock units (Photo B–6). Baker is a good place to gas up and to get supplies before you head north to Death Valley on Highway 127.

Photo B–4. Foreground, Soda playa south of Baker with dust clouds whipped up by strong north wind. Light band in midground is the Kelso dune complex with Granite Mountains forming right skyline. (Photo by Helen Z. Knudsen.)

Photo B–5.
Bumps on the skyline are cinder cones of the Cima volcanic field as seen from I-15 descending into Soda Lake basin. Soda Dry Lake is white area in foreground. (Photo by Helen Z. Knudsen.)

Inbound: Start from the stop sign in the middle of Baker, proceed west on Main Street. Baker Hill is 0.6 mile out on the right side, and the gravel pit is in view 0.3 mile farther. Mileage posts along I-5 are every 0.5 mile on both sides.

● Travelers proceeding to Las Vegas should stay on I-15 and skip to Segment F—Baker to Las Vegas, in Chapter 4.

Photo B–6.
West end of Baker Hill looking north from I-15 approaching Baker. Upper half of hill is darker, cavernously weathered Paleozoic carbonate rock thrust over underlying pre-Paleozoic, more variegated, metamorphosed rocks with light spots. Plane of thrust is nearly flat and halfway up the hill face. (Photo by Helen Z. Knudsen.)

Segment C—Baker to Shoshone, 55 miles

▶ Note odometer reading at Baker.

● A useful navigational aid on this route (Figure 3–7) is county mileage markers (MM), like Figure 3–3, every half mile at the highway's edge. North on Highway 127 in the wide valley toward Shoshone, rocks in the mountain front immediately left are Precambrian metamorphics, probably around 1 billion years old (perhaps more), and much younger Mesozoic (150 m.y.) granitic intrusives. The light spots are outcrops of marble unable to maintain a coating of rock varnish. Rocks in Hollow Hills to the east are also largely Precambrian metamorphics and local Mesozoic igneous intrusives. Within a mile the high Avawatz Range looms on the skyline at 11 o'clock. Large alluvial fans extend west from Hollow Hills and, farther north, east from Avawatz Mountains.

Inbound: Most of these features are seen within 3.5 miles of Baker.

● At 3.5 miles from Baker, the floor of Silver Dry Lake lies close on the left and continues for the next 5.5 miles. This is part of the bed of ancient Lake Mojave which was nourished largely by an expanded discharge from the Mojave River. Even now, in wet winters, Mojave River flood waters get this far. The lake flat was flooded by three feet of water in 1968 and 1969, and in 1916 the basin was filled to a record depth of 10 feet.

Keep watching the base of the hills along the far (left) shore. You should be able to make out disconnected parts of a horizontal wave-cut cliff marking a Lake Mojave water level (Photo C–1). From 6.5 to 7 miles from Baker is a good place to look. Native Americans liked the shore of this lake, and artifacts from their occupation 10,000 and more years ago are found at many sites along this old strand line.

Inbound: Shorelines are well seen when backlit along the western edge of Silver Dry Lake beyond the power lines, especially near MM 8.00.

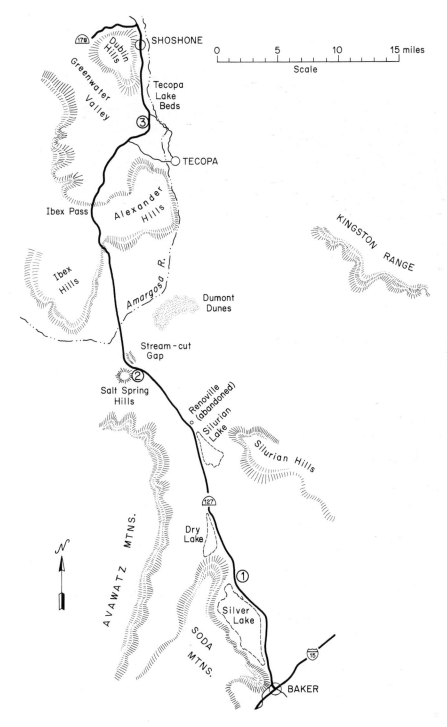

Figure 3–7.
Segment C, Baker to Shoshone.

Photo C–1.
Looking west across Silver Lake playa from Highway 127. Faint horizontal line along base of hills marks ancient Lake Mojave water level.

● Piles of light-colored material near the road, seven miles from Baker, are talc, dumped by trucks hauling from mines to the north. Continuing north look left for shoreline remnants and read ahead in ①.

① Approaching the power-line crossing at the north end of Silver Lake, the horizontal line of the old shoreline is clearly visible in favorable late afternoon lighting along the base of the hills across the lake. The light-colored rock knobs here and at the northwest corner of Silver Lake are marble, composed of the carbonate rock, dolomite, which is like limestone except for its higher magnesium content. Immediately left of the highway, 0.3 mile before you pass under the power lines, is a low beach ridge closing off a shallow basin that lies on its north side. A good view of this relationship is seen if you look back at 7 o'clock from the first power-line crossing. The broad low ridge under the power line is another slightly older beach.

Inbound: Four power lines are crossed 0.5 mile south of MM 10.00, and the dumped talc is a little south of MM 8.50. Shoreline features on the far right side are easier seen inbound.

● Within a mile beyond the power lines, beyond the first curve left, are the Avawatz Mountains at 10 o'clock. Their southern flank, at 9:30 o'clock, is underlaid by the Avawatz Formation, a lower Pliocene-upper Miocene (4 to 8 m.y.) fossil-bearing, terrestrial, sedimentary accumulation lying partly within the complex Soda-Avawatz fault zone. The darker rocks making up most of the Avawatz range front are Precambrian metamorphics.

Inbound: Avawatz Mountains have been to the right ever since the Salt Spring Hills.

● At three to four miles beyond the power line, the Silurian Hills attract attention at 2–2:30 o'clock. They feature an extremely complex structural arrangement of old Precambrian, younger Precambrian, Paleozoic, Mesozoic, and Cenozoic rocks of igneous, metamorphic, and sedimentary nature. Silurian Hills will be in view for a good many miles, and occasional glances to the right will give a sense of their lithologic and structural complexity.

▶ Note odometer at the power lines.

Inbound: Silurian Hills are seen left across Silurian playa and also from farther south.

● After making a second major curve, this one to the right, about four miles from the power lines, look at the large alluvial fans built eastward from Avawatz Mountains at 9–11 o'clock. If light is good, several generations of fan surfaces can be distinguished on the basis of differences in darkness, desert pavement development, and dissection (gullying). Especially good views of these relations are seen at 9:30 o'clock just short of the next curve.

Inbound: Large fans at the base of the Avawatz are well seen coming south from Silurian playa, at, before, and after MM 19.00. Dissection is well seen at 3 o'clock from MM 16.00.

● Beyond this third curve the Silurian Hills are in good view across the valley at 2 o'clock. On the distant skyline at 1–1:30 o'clock, high rugged peaks in the Kingston Range loom. The Kingstons house 7,000 feet of weakly metamorphosed, late Precambrian sedimentary rocks, largely shale, sandstone, quartzite, limestone, dolomite, and conglomerate. These rocks are collectively termed the Pahrump Group, which you see later. The often snowcapped far-skyline mountain behind the north end of the Kingstons is Charleston Peak (11,918 feet) in the Spring Mountains of Nevada. It continues in view for some distance.

Inbound: Kingston Range and Charleston Peak are seen to the left south of Silurian playa.

● About 10 miles from the power lines, Silurian playa lies immediately right of the highway. Now dead ahead, 10 to 12 miles away, are the Dumont dunes (Photo C–2). They consist of sand piled up by winds from several directions. Raw alluvium deposited by the Amargosa River is a principal source of their sand.

Inbound: Silurian Dry Lake lies to the left between MMs 20.50 and 20.00. Dumont dunes have been seen earlier.

● Beyond Silurian Dry Lake the road again curves left, and at 11 o'clock, about five miles ahead, are Salt Spring Hills, composed of Cambrian (500 to 570 m.y.) quartzite beds. When closer, note the two different shades of rock varnish—brown and black. The brown forms on white to pink quartzite layers and the black on very dark quartzite. A good view of the Dumont dunes comes at 1 o'clock from MM 23.0.

Inbound: Salt Spring Hills are on the right between MMs 28.00 and 27.50.

② In another 5 miles (17.3 miles from the power lines), the highway again swings left, and ahead at 12:15 o'clock is a narrow gap through a rock ridge right of the highway (see Photo C–2). This gap was cut by overflow from Lake Mojave in pluvial times when it ran into Death Valley by joining the Amargosa River ahead. Between MM 26.50 and 27.0 pale, soft, fine-grained deposits and some weak shoreline features, such as a low wave-cut cliff and horizontal marks like bathtub rings at the base of the Salt Spring Hills to the left, show that ponding occurred here before cutting of the gap was completed. At the north end of the Salt Spring Hills the highway makes two or three little curves, and when it straightens out green trees at 1 o'clock mark the location of the gap.

Inbound: The gap is seen looking back between MM 26.00 and 27.00.

If you would like to have a closer look at this gorge, a rocky desert road turns off right just short of the little granitic knob left of the highway less than 0.5 mile ahead (19.7 miles from the power lines). The turnoff is roughly 75 feet beyond the Highway 127 sign, and the distance to the gap is 0.3 mile. The road is now blocked part way in, but the additional walk is easy. It is interesting to imagine what this spot was like when it harbored a stream powerful enough to cut a gorge into granitic rocks. Even now, the growth of salt cedar and cane grass make this

Photo C–2.
High-altitude vertical air photo of Salt Spring Hills area, north to top, scale lower right. (U.S. Air Force photo taken for U.S. Geological Survey, 744V-074.)

spot unusual. Right of the gap is a mass of dark, variegated, Cambrian sedimentary rocks, and farther to the east-northeast are exposures of the same Cambrian quartzites seen in Salt Spring Hills.

Inbound: Road into the gap is just beyond MM 29.00.

● Back on Highway 127 and rounding the corner, you get a good view of granitic rocks at 3 o'clock and of the Cambrian quartzites at 2 o'clock. Large boulders left of the highway were carried from the Avawatz Mountains by debris flows. The southern end of the Death Valley depression lies ahead to the left. A dirt road takes off into it within a mile, at the Harry Wade Historical Monument identified by a Historical Landmark sign. You will enter from the side some 30 miles farther north. Small sand dunes are seen right of the road at 1 o'clock from this intersection (see Photo C–2). You come to the Amargosa River in 2 miles, marked by a line of salt cedar trees and a sign. Don't be surprised to see water running here in winter or spring. The river rises in the high Spring Mountains of Nevada and at times flows all the way to the salt flats of central Death Valley.

Inbound: Amargosa River is crossed just beyond MM 32.00. The small dunes are at 31.50 and the road in is at 31.00. Harry Wade Monument and Death Valley Road are just beyond MM 30.00.

● Beyond the Amargosa River, the southern end of the Ibex Mountains (Saddle Peak Hills) lies at 10–11:45 o'clock. The dark reddish-brown rocks there are part of the late Precambrian Pahrump Group of sedimentary beds, and the lighter rock capping the tops of ridges at 11:30 o'clock is the slightly younger Noonday dolomite.

Inbound: You need to look back from near MM 33.00 for this view, but you pass close by these rocks on the straight part of the run down from the Ibex Pass.

● In another mile the Dumont sand dunes are in view at 3:15 o'clock, and a well-graded road to them turns off right shortly. This makes an interesting side trip but be careful not to get stuck hub deep in soft ground near the dunes.

▶ | Note odometer reading at the Amargosa River crossing.

Inbound: The turnoff is near MM 34.00.

● At 5.5 miles from the Amargosa River, the Noonday dolomite is close by on the left. The subdued gray hills at 1-3 o'clock are composed largely of fanglomerate. In places this deposit contains lenticular beds of large angular fragments of a single type of rock, which probably represent landslide and rock-fall accumulations. Because they contain just one kind of rock, they are called *monolithologic breccias.* This fanglomerate is similar in character, and possibly equivalent in age, to a widespread late Pliocene-early Pleistocene (2 to 4 m.y.) deposit of the Death Valley area, the Funeral Fanglomerate.

Inbound: The Noonday exposure is about 3 miles from the relay station near MM 36.50.

● About 7 miles beyond Amargosa River a microwave relay station left of the highway is passed, and the ascent of the Ibex Pass begins. The dark rocks east of the highway are volcanics. Deep road cuts 1.5 to 2 miles up the grade are in badly fractured granitic rocks, but nearer the Inyo County line you pass into uplifted fanglomerates containing largely cobbles of granitic and volcanic rocks.

▶ | Note odometer at the summit just beyond the Inyo County line.

Inbound: Relay station is 1.8 miles south of Ibex pass, near MM 39.50.

Mileages on highway posts begin again at zero in Inyo County, with 0.5 mile spacing, full miles on the right, and 0.5 miles on the left side. As you descend from Ibex Pass, the northern Ibex Mountains—largely a complex of old and late Precambrian rocks—are on the left. White spots are talc workings. The well-bedded rocks of complex structure at about 10:30 o'clock are Cambrian. The variegated peak at 10:45 o'clock at the north end of Ibex Mountains is Sheephead Mountain. Within 1.5 miles from the pass, you get good views of the dissected badlands formed in pale Tecopa lake beds on the valley floor at 11–1 o'clock. Our old friends the dark Cambrian quartzites of Salt Spring Hills are seen in the near ridge at 1 o'clock.

The high ranges beyond the lake basin appear striped because of layering within the thick section of early Paleozoic sedimentary formations composing them. The nearer range is the Resting Spring, and the far skyline range is the Nopah. The combined thickness of the Paleozoic beds exposed in these mountains is 23,000 feet, ranging from Cambrian (570 m.y.) to Pennsylvanian (300 m.y.).

Inbound: To get this view, ascending toward Ibex Pass, you will have to stop near MM 3.50 and look back.

● In another mile the Dublin Hills, with beautifully layered Cambrian sedimentary formations, loom up dead ahead. The wide valley at 11 o'clock is Greenwater Valley, which you enter from the side farther north.

Inbound: Look right from MM4.

● In another mile you are within the low hills of dissected and rilled Tecopa lake beds. A thin layer of younger, dark, alluvial gravel laid down on top of the soft lake beds, before dissection, drapes locally over the slopes, masking the deposits beneath. Where the lake silts have good coherence, steep castellated cliffs have formed.

Inbound: Near MM 5.

③ Just a bit more than 2 miles beyond the first paved turnoff to Tecopa, low, crumbled remains of adobe walls of the old Amargosa borax works are close to the highway on both sides. This site was used during summers from 1882 to 1890, when temperatures on the floor of Death Valley prevented crystallization of solutions at the Harmony borax plant.

Inbound: Ruins of the Amargosa Borax works are 7 miles south of Shoshone, 0.9 mile beyond the road to Hot Springs and Tecopa.

● Starting about 0.3 mile beyond the second paved turnoff to Tecopa, you may be able to see faint traces of old, narrow, hand-dug trenches extending down spurs close to the road on the left. Don't be confused by fresher trenches and much wider bulldozer scars. These old, largely infilled trenches, about 2 feet wide and now only a foot or so deep, were dug during World War I in search of nitrate deposits, the supply from Germany having been cut off. It is amazing that they have survived so long. Approaching Shoshone the power line east of the highway follows the old Tonopah-Tidewater Railroad right of way.

Inbound: As you go south from Shoshone watch for hand-dug trenches on spurs right of the highway near MM 10.50. Ruins of borax plant are 7 miles from Shoshone.

● A little side trip just 0.2 mile short of Shoshone is well worthwhile. Turn left at the 127/178 (Charles Brown Highway) intersection onto a two-wheel track dirt road, which forks in about 100 feet. Keep left straight ahead for 0.2 mile to a 20 foot high cliff exposing white volcanic ash in which caves have been excavated. The ash bed is about 10 feet thick, muddy in its upper part, and well bedded there, suggesting it has been at least partly reworked from surrounding slopes. You may be surprised to learn it came from a huge volcanic explosion in Yellowstone Park 600,000 years ago. It is the famed Lava Creek ash. Ash beds are useful as time reference horizons. Continue a few hundred yards on up the road to see the large quarry pits in the ash.

Inbound: Turn off 127 to the right opposite the 178 separation to the east.

● Shoshone is the point of departure to Death Valley; food, gasoline, water, and a motel are available. The right of way of the old Tonopah-Tidewater Railroad is just to the east at "*Shoshone Population 100*" near the north edge of town.

Segment D—Shoshone to Death Valley Floor at Ashford Mill via Salsberry and Jubilee Passes, 29 Miles

● The drive from Salsberry Pass to Furnace Creek Ranch is magnificent; plan to do it at a leisurely pace. In the first mile going north from Shoshone on Highway 127 (Figure 3–8), the black rock chunks on slopes immediately to the left are derived from the Funeral basalt lava flows, 4 m.y. old. The 500 foot high ridge of well-layered pink and brown rocks two miles to the right consists of Plio-Miocene (3 to 8 m.y.) volcanic and sedimentary deposits. It is part of a fault block lying in front of the darker Cambrian beds (500 to 570 m.y.) of the Resting Spring Range. Approaching the Salsberry Pass turnoff (Highway 178) about 1.5 miles north at "*55 Miles to Badwater,*" the abandoned Gertsley borax mine (white spot) is visible near the base of the hills at 2:45 o'clock.

▶ | Note odometer reading at turnoff.

Inbound: Pause at 127/178 intersection to see these features.

● Proceeding west on Highway 178, you circle the north end of Dublin Hills. In about 1.5 miles Greenwater Range fills the skyline from 11:30 to nearly 3 o'clock, the part seen here

being largely Tertiary volcanics. The distant sharp peak at 3 o'clock is Eagle Mountain, on the Death Valley Junction road. It is composed of Cambrian sedimentary beds.

Inbound: This location is near MM 40.

● Within another mile you begin to see that the east flank of Greenwater Range is locally mantled by a thin layer of black rock tilting eastward that rests on more highly colored, older Tertiary volcanics. These black rocks are flows of the Funeral basalt, which you see scattered throughout the Death Valley country.

Inbound: This view is seen from vicinity of MM 39.

● About five miles after the turnoff you are descending gently into wide Greenwater Valley. Ahead at 11 o'clock is Sheephead Mountain, composed of jumbled, colorful volcanics. In another two miles you will be near the center of the valley, a broad synclinal down-warp. You earlier crossed its southern end on Highway 127 between the two turnoffs to Tecopa. To the north, dark layers of Funeral basalt mantling both its sides are inclined inward toward the valley axis. This is evidence for synclinal downwarping of geologically recent date, because the Funeral basalt is no more than 4 m.y. old.

Inbound: MMs 36 and 37 are in center of Greenwater Valley.

● In two more miles the highly irregular color pattern in the hills at 1:30–2:30 o'clock suggests a complex mixture of rocks. Those exposures are indeed part of a structurally jumbled mass, appropriately named the Amargosa chaos. More specifically, they belong to the Calico phase of the chaos, an obviously fitting name. Calico rocks have bright, varied colors in mixed patterns. Other, less colorful phases of the chaos are the Virgin Spring and Jubilee, to be seen ahead. In simplest terms, the chaos phases are breccias, tectonically shattered and jumbled blocks or sheets of rock. Landslides and rock falls occurring from freshly created steep slopes probably contributed to the complexity of the chaos deposits.

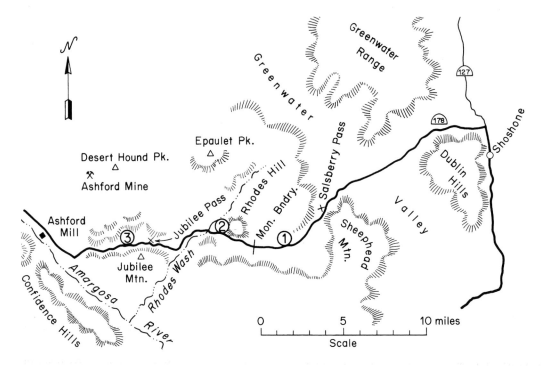

Figure 3–8.
Segment D, Shoshone to Ashford Mill (Death Valley).

● A complex of Tertiary volcanic flows and tuffs is seen south of the road ascending to Salsberry Pass (3,315).

▶ Note odometer reading at the summit.

Inbound: Summit of Salsberry Pass is four miles east of the Monument Boundary. Calico chaos is near 8 o'clock, two miles east of the summit at MM 34.00.

① Within a mile after crossing Salsberry summit, you pass onto a fairly smooth alluvial surface. Road cuts show these gravels to be whitened by secondary deposition of calcium carbonate (*caliche*). About 1.5 miles farther, roughly that distance beyond "*Elevation 3,000 ft.,*" you might stop to survey the country, be-

cause geologically it's a little complex. The dull, dark rocks in the near hills to the left at 9–10 o'clock are old Precambrian metamorphics. The closest rugged dark rock mass at 1 o'clock (Rhodes Hill, see Figure 3–8) is Precambrian gneiss. The more colorful outcrops near its northern base are part of the Virgin Spring phase of the Amargosa chaos in fault contact with the gneiss. At 2–2:30 o'clock on slopes farther right are highly colored rocks of the Calico chaos. The smooth-topped skyline hill at 2:15 o'clock (Epaulet Peak) is capped by our friendly Funeral basalt. The highly colored rocks of sharp, skyline Salsberry peak at 3:15 o'clock are also part of the Calico chaos. At the Monument boundary mileage posts change again (Figure 3–9) and Highway 178 ends.

Inbound: The view point is 1.5 miles outside the Monument Boundary.

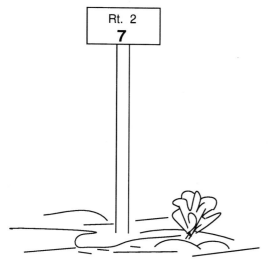

Figure 3–9.
Death Valley National Monument mileage
marker (MM).

● At "*End 178*" across the wash at 10–11:30
o'clock is a nice outlying knob of Virgin Spring
chaos resting on somber old Precambrian schist.

② At 1.6 miles inside the Monument Boundary,
stop between the little rock knobs just right of
the highway. The greenish-gray somber ridge-
like knob at the east end of the group is old Precambrian schist. The other more colorful and
heterogeneous knob to the west is made of Virgin Spring chaos breccias of various rock types.
At 2:45 o'clock on the skyline is Epaulet Peak,
and on its near slopes is Calico chaos. The
darker but variegated rocks, still lower at 2:30
o'clock, are the Virgin Spring chaos. They rest
on some uniformly dark, greenish-gray, old
Precambrian rocks at 2 o'clock, which are in
the core of an anticline. Low down at 1 o'clock
is some more Virgin Spring chaos on the south
limb of the anticlinal structure. The highly colored rock knob alongside the road just ahead is
also Virgin Spring chaos. Scattered knobs and
patches of the same chaos resting on old Precambrian rocks are seen just left of the wash as

you proceed down the highway. In the Monument distances are indicated by mileage markers
like Figure 3–9.

*Inbound: This locality is 0.4 mile east (beyond)
MM 2.*

● In 0.9 mile, across the wash just left of the
highway, is a rocky knob displaying much cavernous weathering. If you stop and walk over,
you will find it is composed of broken up (brec-
ciated) carbonate rock (dolomite), probably of
Cambrian age. It is part of the Virgin Spring
chaos and rests in fault contact with somber
early Precambrian rocks on its west side. In this
region, cavernous weathering is a characteristic
of chaos rocks.

Inbound: This knob is 1.5 miles east of MM 4.

● In just a bit more than 7 miles from Salsberry Pass, where the road curves south (left),
is a high reddish cliff close on the left side.
About 30 feet above its foot chaos rests on
weathered, rust-stained, Precambrian rocks.
The contact is one of the many low-angle fault
surfaces of the area.

*Inbound: Cliff with fault is 1.3 miles from the
crossing of Rhodes Wash.*

● Around the curve and beyond the cliff, the
subdued near hills at 2–3 o'clock are composed
of Funeral fanglomerate, a Pliocene (4 m.y.)
deposit that is younger than the chaos. The
high, pointed, dark peak ahead at 1 o'clock is
Jubilee Mountain, composed of a coarse-
grained, old Precambrian gneiss.

Directly down Rhodes wash you see the floor of
Death Valley and the rounded whale-back surface of Confidence Hills, a faulted anticlinal
structure in soft late Tertiary sedimentary rocks.

*Inbound: Rhodes wash is crossed 1.2 miles
from Jubilee Pass.*

● Shortly the highway crosses Rhodes wash,
curves west, and ascends for a mile to the summit of Jubilee Pass. From here the high point
on the skyline at 2 o'clock is Desert Hound

Peak. The large area of variegated, brown to reddish rocks on its lower slopes toward you and to the west is all Virgin Spring chaos. The upper part of the peak consists of old Precambrian metamorphics that compose the core of the Desert Hound anticline that extends north to Mormon Point.

> ▶ Record odometer reading here.

Inbound: Summit of Jubilee Pass is 4.6 miles from road junction on Death Valley floor.

③ In about 1.4 miles from Jubilee Pass near MM 7, close to the road on the north, are three prominent knobs of Jubilee chaos with typical cavernous weathering. The ridge just beyond, left of the highway, exposes a series of tilted red sandstone, fanglomerate, and volcanic layers of Tertiary age. The smooth pink and white areas seen 0.25 to 0.50 mile left of the highway on this ridge are volcanic tuffs.

Inbound: Jubilee knobs are on the left side approaching MM 7.

● Beyond the red sandstone-fanglomerate ridge and just right of the highway, the knobs with good cavernous weathering consist of Jubilee chaos, but the last big rock knob left of the road, about 3.5 miles from Jubilee Pass, consists of limestone and dolomite breccia in the Virgin Spring chaos.

Inbound: Virgin Spring knob is 0.5 mile beyond MM 10 on right.

● About four miles from Jubilee Pass near MM 8 the southern part of Death Valley is fully in view. Confidence Hills are at 9–12 o'clock, and Shoreline Butte is at 2 o'clock. The horizontal shoreline markings on its slopes become more apparent after you turn right (north). They are best seen in late afternoon light. A sharp eyed observer should make out at least a dozen levels.

Inbound: Confidence Hills are at 4 o'clock going up Jubilee Pass Road.

● These strand lines were cut by pluvial Lake Manly, at least 600 feet deep and well more than 100 miles long, which lay in Death Valley between 10,000 and 75,000 years ago. It was fed by a greater discharge from the Amargosa-Mojave rivers system and by water that flowed through Wingate Pass (see Photo 2–21) into Death Valley from a deep lake in Panamint Valley that was fed largely by runoff from the Sierra Nevada (*see* Basin Ranges, province, special features).

Inbound: Shorelines have already been seen north of Ashford Mill.

● As you turn right (north) at the road intersection, 4.7 miles from Jubilee Pass, the steep front of Black Mountains lies to the right. This is a fault scarp with large patches and knobs of Virgin Spring chaos on its face and along its base as far north as Ashford Canyon (at 1:30 o'clock). Beyond there the scarp consists of old Precambrian rock composing the west flank of the Desert Hound anticlinal core.

Inbound: Intersection is 2 miles south of Ashford Mill. You turn left (east) on what will become Highway 178.

● The ruins of Ashford Mill lie left of the road two miles north from this intersection. The valley floor here is windswept as shown by little lee-side sand tails behind bushes and large rocks, and the wind pitting, fluting, and polish on stable surface stones.

Inbound: Seen as you drive south from Ashford Mill.

Segment E—Ashford Mill to Furnace Creek Ranch, 44 Miles

> ▶ Note odometer mileage opposite Ashford Mill turnoff, figure 3.10.

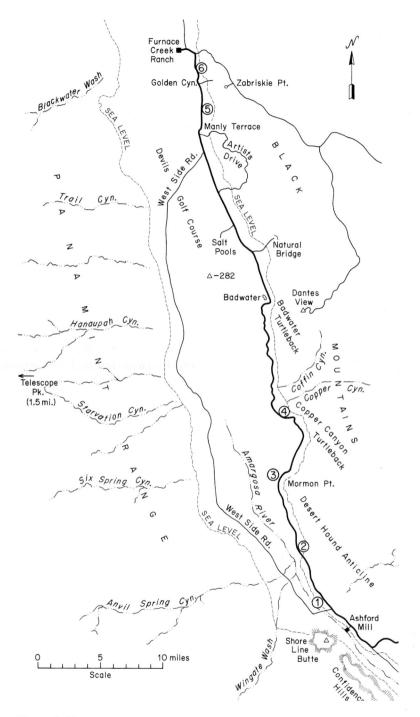

Figure 3–10.
Segment E, Ashford Mill to Furnace Creek Ranch.

● Watch for mileage markers (MM) along this route. Starting 0.5 mile to the north, three Funeral basalt knobs partly mantled by wind-blown sand rise above the fan surface just left of the highway. Their north-south alignment indicates that they mark the trace of a fault block uplifted on its west side. To the right, forelying rocky knobs scattered outward from the base of the mountains at 2–3 o'clock consist of Jubilee chaos. Many horizontal shorelines cut into the northeastern face of Shoreline Butte are visible in the first mile or two north from Ashford Mill. Big, wide open Wingate Wash in the Panamints comes into view at 9 o'clock as you clear Shoreline Butte.

Inbound: The first basalt knob comes on the right 0.2 mile south of West Side Road.

● In two miles West Side Road (dirt) takes off. This is a good place to stop and look around. At 10:30 o'clock near the center of the Death Valley floor is a small cinder cone located on a branch of the Death Valley fault zone. On the 9 o'clock skyline is Wingate Pass (see Photo 2–21) through which water flowed perhaps as long as 75,000 years ago and as recently as 10,000 years ago, from the large pluvial lake in Panamint Valley. Immediately right of the road a fault scarp in black Funeral basalt extends north for more than a mile. In places Funeral fanglomerate, rather than basalt, composes the scarp. It marks the trace of another fracture within the Death Valley fault system.

Inbound: This fault starts three miles south of MM 33 and ends at West Side Road.

① In 3.2 miles north from Ashford Mill at a curve in the highway, with good light, one can see how the little cinder cone, now at 9 o'clock, is sliced apart by right-lateral faulting.

Inbound: About two miles south of MM 33 at 1 o'clock.

● About 4 miles north of Ashford Mill, at 1–3 o'clock, is a large much-dissected body of Funeral fanglomerate at the base of the Black Mountains. At MM 32 the highway curves east toward Black Mountains, and the contact between fanglomerate and early Precambrian gneiss composing the mountains is seen to be the Black Mountains frontal fault. This is a zone consisting of individual linear fractures steeply inclined to the west, but with segments locally diverging from a predominate north-northwest bearing. Individual fractures within this zone have recently been active, and you will see fresh scarplets breaking many fan surfaces along the mountain base. Some are probably no more than a few hundred to a few thousand years old.

Inbound: About 0.6 mile south of MM 33.

● Now is a good time to compare huge alluvial fans on the west side of Death Valley with the much smaller east-side fans (Photo E–1) over which you drive. This difference partly reflects the greater amount of water and rock debris discharged by the larger canyons of the high Panamint Range, but it is also a result of eastward tectonic tilting of the Panamint-Death Valley block. Geological relationships suggest that much of this tilting has occurred recently. Tiltmeters and surveys indicate that it continues today. For this reason the Death Valley salt pan lies close to the Black Mountains in many places including Badwater. The effect is similar to tilting a saucer partly filled with water.

Tilting has caused fans on the west side of Death Valley to increase by extending outward onto the valley floor. At the same time, tilting depresses fans on the east side causing them to become partly buried by valley-floor deposits, thus reducing their size. The Black Mountains frontal fault defines the eastern edge of the tilting block, and fault scarplets breaking Black

Photo E–1.
Small alluvial fans at base of Black Mountains north of Coffin Canyon. (Photo by John S. Shelton, 3441.)

Mountain fans are related to this movement. Death Valley is a structural depression; its form has been determined by deformation, probably both warping and faulting. Some geologists call this part of Death Valley a pull-apart basin. They think of it as a rhombohedral shaped block that dropped down because of the extension occurring between two parallel strike-slip fault zones. Any closed depression such as Death Valley is geologically suspect. Nature abhors a vacuum, and she despises closed depressions. She fills them with anything available. In humid areas the initial filling is water, succeeded later by sediment. In dry regions sediment is the principal filling. Even though parts of Death Valley are underlain by 3,000 feet of young alluvium resting on top of 6,000 feet of Tertiary sediments and volcanics, nature has not yet been able to completely fill this depression. It has to be very young and has formed rapidly.

With an elevation 282 feet below sea level, Death Valley is the sump for a large part of southeastern California and adjacent areas of Nevada. If today's climate were more humid, streams would run to Death Valley from all directions, just as they did in pluvial times, and it would harbor a large lake, which could be over 2,000 feet deep before spilling over its rim. Because of its depressed level, Death Valley is much better "watered" by springs and streams (Furnace Creek) than most of the surrounding desert.

Inbound: In the vicinity of MM 33.

② The west face of Black Mountains is a youthful fault scarp, locally characterized by wineglass canyons; a good example is just ahead at nearly 7 miles north of Ashford Mill and 0.7 mile beyond MM 33 (Photo E–2). The base of the wineglass is the spreading fan at the foot of the mountain, its stem is the narrow steep-walled gorge cut through the mountain front, and the bowl is the open area of dispersed headwater tributaries.

Inbound: The wineglass canyon is a mile south of MM 35. The best view is looking back.

This reach of the mountain front is made up of Precambrian gneiss and carbonate rocks that yield large fragments. The fan surfaces are therefore rough, irregular, and dotted by good-sized boulders. At and just north of the wineglass canyon, watch the toes of the fans where they come close to the salty flats near the highway. Many of the fan boulders have a decrepit appearance because they are being disintegrated by the growth of salt crystals within their pores.

Inbound: Examples of disintegrated boulders are near MM 35.

● About 7.5 miles north of Ashford Mill the road curves back toward the mountain front. Directly ahead is a small alluvial cone with a steep bouldery surface displaying patches with different shades of rock (desert) varnish, from gray to dark brown. This and similar cones along the base of Black Mountains have been built largely by rocky debris flows and are best termed debris cones. Flows have inundated parts of this cone at widely separated intervals, as indicated by the variations in shades of varnish.

The grayish lobe in the south central part marks the most recent flow, possibly less than 100 years old. Note the old, faint, game or Native American trail that crosses the lower part of the fan. It is best preserved in the older, more heavily varnished parts of the cone's surface and has been obliterated by the most recent flows.

Inbound: This alluvial cone is about 0.5 mile south of MM 35. You may have to look back to see it.

● About 9 miles from Ashford Mill, 0.7 mile beyond MM 35 and just beyond a place where the road is crowded against the mountains by a salt pond, is a little fan with a fault scarp 7 to 8 feet high breaking its surface. The scarp parallels the mountain front about 20 to 30 feet from its base. Watch for similar scarplets crossing

Photo E–2.
Wineglass canyon in western face of Black Mountains nearly seven miles north of Ashford Mill, beyond MM 33.

fans on up the valley; another one is about 1.5 miles ahead near MM 37. Coarse, moderately cemented fan gravels adhere to the base of the mountain front north from here. They have been relatively elevated by movement on the Black Mountains frontal fault and are sharply incised by gullies from the mountains.

Inbound: These features are seen near and for over a mile beyond MM 37.

● Left at 9:30–10 o'clock across the valley, you see that the upper parts of the Panamint fans are dissected. This is a result of eastward tilting that steepens the fans and causes streams to cut down.

③ At almost 11 miles from Ashford Mill, 0.2 mile beyond MM 37, is Mormon Point (small sign on left). It is at the northern end of the Precambrian core in the Desert Hound anticline (Photo E–3) seen earlier from Jubilee Pass. At Mormon Point the range front shifts to the right because of complexities in the frontal fault system. A brief stop here for a look northward

along the face of the mountains is worthwhile. Note particularly the change in its character as Precambrian rocks give way to less resistant, heterogeneous, and highly colored Tertiary volcanic and sedimentary rocks.

Inbound: Just beyond MM 38.

▶ | Record odometer reading at Mormon Point sign.

● As the road curves right toward the mountains, 0.5 mile beyond Mormon Point, a mass of Funeral fanglomerate straight ahead lies in fault contact with old Precambrian schist, gneiss, and marble, which compose the high mountain face behind. Younger fan and lakeshore gravels locally mantle the north slope of the Mormon Point peninsula (Photo E–4). Keep your eyes peeled for small fault scarplets cutting across alluvial fans all the way around the Mormon Point reentrant, especially in the eastern part.

Photo E–3.
View north up central Death Valley from south of Mormon Point and Desert Hound anticline. (Photo by John S. Shelton, 4238.)

Photo E–4.
Horizontal lake shorelines cut in gently inclined Pleistocene fanglomerate deposits east of Mormon Point, view to south. (Photo by John S. Shelton, 3459.)

Inbound: Fanglomerates and faults are to the left between MMs 39 and 38.

● About 1.5 miles from Mormon Point, the road straightens into a north-northeast course. Here you look dead ahead to the southwest limb of the Copper Canyon turtleback (Photo E–5). The nose of the structure is at the Precambrian-Tertiary contact ahead at about 11:30 o'clock.

A turtleback is an unusual geological structure. It consists basically of a mass of Precambrian rock in the core of a plunging anticline that has been exposed by erosion of the overlying deposits. Topographically it has some resemblance to the carapace of a turtle. Where remnants of the overburden remain, they are in fault contact with the Precambrian core of the fold. According to some interpretations, faulting occurred before the folding; according to others, it is considered essentially a contemporaneous event. The turtleback itself is composed of the Precambrian rocks underlying the anticlinally folded fault surface. These structural relationships are best seen a little south of the mouth of Copper Canyon ahead. You view another turtleback beyond Badwater. There is a Mormon Point turtleback too (see Photo E–3), but its relationships are not so clearly evident from the road.

Inbound: You start this straight stretch in reverse at 0.4 mile beyond MM 40.

● Fans with scattered large boulders, 2.6 miles beyond Mormon Point, have been inundated by debris flows that are powerful transporting agents. The bouldery fans along the northeast straightaway are composed largely of old, hard Precambrian gneiss fragments.

● At 3.8 miles from Mormon Point, where the road starts to curve left around the first of several small alluvial fans south of the huge Copper Canyon fan, look to the base of the mountains at 3 to 4 o'clock to see several small debris-flow cones that display surface patches with strikingly different degrees of rock varnish

Photo E–5.
Copper Canyon turtleback viewed from west. (Photo by John S. Shelton, 4241.)

Photo E–6.
Debris cone displaying four degrees of rock (desert) varnish, reflecting incremental building of cone.

(see Photo E–5). One cone has patches with four degrees of varnish (Photo E–6). The horizontal lines on the mountain face a little above the heads of these cones are traces of Lake Manly shorelines.

Inbound: This stretch starts about 2.5 miles south of MM 43.

● At 4.2 miles from Mormon Point is room for parking on the left. This is a good place to study salt weathering of rocks on either side of the road. As salt water soaked up by these rocks evaporates, growing salt crystals disintegrate the rocks. Little piles of rock chips here were once large boulders of solid rock (Photo E–7). Layering or lamination in any boulder is exploited by the salt. This location is about opposite the third varnished debris cone.

Inbound: At 2.6 miles south of MM 43.

④ Stop at less than 0.5 mile beyond MM 42 on the reddish to pink gently sloping Copper Canyon fan and look around. In the mountain front to the east are gray Precambrian metamorphic rocks in the core of the Copper Canyon turtleback (see Photo E–5). They are overlain by brownish and reddish beds of the Miocene (about 6 m.y.) Copper Canyon conglomerates. When you continue north, if the light is right you will be able to see layering in these conglomerates dipping directly into the Precambrian rocks. This indicates a fault relationship between these units.

The narrow slot of the mouth of Copper Canyon, at 2 o'clock, is partly obscured by a degraded fault scarp about 75 feet high in gray fan gravels. A look back toward Mormon Point should show, in reasonable light, lake shorelines cut across the fanglomerates

Photo E–7.
Piles of dark angular rock fragments are remnants of boulders shattered by salt weathering.

(see Photo E–4). Before you get to the apex of the fan, the plunging nose of the Copper Canyon turtleback comes into view. North from Copper Canyon, rocks in the Black Mountains change from somber Precambrian metamorphics to variegated Tertiary sedimentary and volcanic units. Dissection of the Six Springs and South Johnson fans left across the valley at the base of the Panamints is striking. Light-colored spots higher on the Panamints are talc mines.

Inbound: These observations can be made about 0.5 miles beyond MM 43.

● At 0.5 mile beyond MM 44 the highway heads westward again as it starts around the large Coffin Canyon fan. On the skyline at 11:30 o'clock is Telescope Peak (11,048 feet), often snowcapped in winter. It is composed of late Precambrian, weakly metamorphosed sedimentary rocks. Telescope Peak is not the highest peak in the United States, but it is one of the tallest, 11,048 to –282 feet.

Inbound: Seen about 6 miles from Badwater and in another 0.3 mile you are onto Coffin Canyon fan near MM 45.

● From here to Badwater the road swings out and in over five small alluvial fans, some so perfect in form as to look almost artificial (see Photo E–1). Their size varies with the area of drainage within the mountains and the ease with which the bedrock therein is eroded. Their steepness reflects the coarseness of debris (stone size) composing them. Copper and Coffin Canyon fans have gentle slopes largely because the Tertiary rocks in their drainage basins yield much fine material. Between fans the salt pan crowds close to the mountains. Nearer to Badwater you cross rough, steep, bouldery fans derived from tough Precambrian metamorphic rocks.

Between the little fans, where the highway closely approaches the mountain front, you see that the rocks within the Black Mountains frontal fault zone are badly chewed up. They are a mess.

Inbound: The small fans, salt flats, and ground-up rocks are 4.6 to 5.9 miles from Badwater.

● In 0.4 mile beyond MM 47, at 2 o'clock, a rock slide from the steep mountain front projects into the valley. You should be able to recognize it as a jumble of huge blocks on the lower part of the mountain face and valley floor (Photo E–8). The slide is probably just a few thousand years old at most. Precambrian rock comes back to the mountain face a little to the north.

Inbound: Slide is 3.6 miles from Badwater.

● If you are traveling in the late afternoon, look left across the valley to the base of the huge Hanaupah fan from the Panamints to see the shadowed face of a long, low fault scarp at

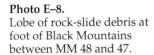
Photo E–8.
Lobe of rock-slide debris at
foot of Black Mountains
between MM 48 and 47.

about 11 o'clock cutting across its lower part (Photo E–9). Don't mistake lines of vegetation for the scarp.

Inbound: From about 3.4 miles south of Badwater to well north of there, the scarp is in view.

● In another mile, remnants of fan surfaces at two levels behind a fault scarp, at the mouth of a canyon to the right, indicate at least two episodes of uplift on the Black Mountains fault. A mile beyond in a similar setting, gravels on a single uplifted surface are heavily covered with rock varnish, which indicates that they have long been undisturbed compared to gravels on the present fan. Soon the Badwater turtleback begins to make the skyline profile ahead on the right.

Inbound: The heavily varnished gravels are 1.8 miles south of Badwater near MM 49.

● By this time you have traveled as much as 15 miles from Mormon Point, and you are probably aware that the wide salt flats hug the base of Black Mountains. This occurs because eastward tilting of Death Valley is relatively strong here. The character of this salt pan surface changes from time to time depending on the water supply. During relatively dry periods it has an irregular pattern of polygonal plates 5 to 15 feet in diameter with turned up edges (Photo E–10). They look a little like the ice floes in classical paintings of Washington crossing the Delaware. These plates appear to grow at their edges by crystallization of salt in cracks. After a severe wetting the salt pan is much smoother.

The breakup of boulders at the toes of fans by salt-crystal prying is well seen just east of the road in this reach. At *"Badwater 1/2 mile"* Hanaupah fan and its fault scarp are in good view left across the valley. The fan you are crossing is also faulted (Photo E–11).

Inbound: About 0.7 mile south of Badwater is a good place to walk out onto salt pan.

Photo E–9.
Fault scarplet, up to 50 feet high, breaks surface of Hanaupah fan at east base of Panamint Mountains. (Photo by John S. Shelton, 4244.)

Photo E–10.
The salt pan after an extended dry period, a little south of Badwater. Polygonal plates grow at edges and push against each other.

● Badwater is worth a stop. Northward on the mountain front at about sea level (*see* sign on the mountain side) are small remnants of cemented gravel adhering to the steep rock face. At least some of them mark the shoreline levels of a pluvial lake. Because Lake Manly was at least 600 feet deep, they represent a lower stage when water depth was only about 300 feet. Small remnants of well-worn shoreline gravels lie at higher levels but are not recognizable from the road. By walking out on the salt pan at Badwater, you see these relationships more clearly. The west flank of the Badwater turtleback is sliced off longitudinally by the Black Mountains frontal fault, and the nose of the turtleback is better viewed a bit farther north. Fans near Badwater are broken by fault scarplets, especially the fan to the south (see Photo E–11).

▶ Before you leave Badwater note your odometer reading.

Photo E–11. First Black Mountain fan south of Badwater showing fault scarp at head and fault graben along highway lower left. (Photo by Ronald I. Dorn.)

Inbound: Badwater is 0.8 mile from the Telescope Peak sign.

● Leaving Badwater you move north along the west flank of the Precambrian core of Badwater turtleback. Within the first mile, considerable salt disintegration of stones has occurred just east of the road. Remnants of gravels cemented by calcareous deposits are seen at about sea level on the mountain face. At *"Telescope Peak, 11,049 feet,"* is a good place for viewing Hanaupah fan across the valley. Much of the dissection of the fan's lower reach originates at the fault scarp.

● In one to two miles from Badwater the turtleback's nose is more apparent at about 2 o'clock. The intimately rilled light-colored rocks there are Tertiary deposits lying above the anticlinally folded turtleback-fault sur-face. Looking to about 3:15 o'clock you should see a little conical knob of reddish-brown Tertiary rock resting on the crest of the turtleback. It is a remnant of the turtle-back's former covering. In the vicinity of MM 53, the large boulders on the fan surface were transported by debris flows.

Inbound: This vista is well seen about 1.6 miles from Devils Golf Course Road, 0.3 mile beyond MM 54.

● In 3.5 miles from Badwater is the turnoff to Natural Bridge, a 50-foot span in sedimentary rocks of the Tertiary Artists Drive Formation. Look back from here for a good view of the Badwater turtleback. From here north, the mountains are composed of Tertiary volcanic and sedimentary rocks aggregating a thickness of 13,000 feet.

The Black Mountains contain a surprisingly small amount of *in situ* Paleozoic rock, considering the tens of thousands of feet of Paleozoics exposed in adjacent areas. It appears that the Black Mountains block, lying between two large fault systems—the Death Valley on the west and the Furnace Creek on the east—has been stripped of its former thick cover of Paleozoic beds either by erosion or tectonic denudation (faulting). Erosion would require about 25,000 feet of uplift, but that is not out of the question judging from paleobarometric indicators in some Black Mountain rocks suggesting they once lay 30,000 feet deep.

● North of Natural Bridge are the beginnings of a group of low hills lying in front of the main Black Mountains block. They consist largely of variegated Tertiary volcanic rocks and lesser amounts of sedimentary deposits, all highly deformed. These hills are part of the subsidiary, fault-bounded Artists Drive block, lying between the Black Mountains block and the Death Valley block (Photo E–12).

Inbound: Natural Bridge is two miles south of Devils Golf Course Road.

● About 4.5 miles north of Badwater the highway starts across an alluvial fan of notably gentle slope. Although large stones dot its surface, especially near MM 55, the fan contains much fine material derived from the soft rocks composing the Artists Drive block. This is the principal reason for its gentle slope and smoothness.

Inbound: Big boulders begin to appear on the fan surface 1.3 miles beyond Mushroom Rock.

● In less than a mile is the turnoff to Devils Golf Course (salt pools). A visit will give you a look at the extremely rough surface of a relatively pure salt pan. This microrelief is a product of solution plus localized crystallization along fractures that forms ridges and spires of solid salt. Yes, it would be a miserable place to play golf; let the Devil have it.

Inbound: Golf course turnoff is 2.3 miles south of Artists Drive entrance.

● In another 2.3 miles is Artists Drive entrance. This is a one-way, 9.5-mile road through an intimately dissected badland with narrow passages cut into highly colored sedimentary and volcanic rocks of the Funeral fanglomerate and Artists Drive Formation. The exit from Artists Drive is 3.5 miles north.

Just left of this turnoff are two low ridges of Funeral fanglomerate faulted or warped up by stresses related to lateral displacement along fractures of the Death Valley fault system. If you feel a need to stretch your legs, walk to the top of the ridge opposite the turnoff. Here you will find vesicular volcanic boulders deeply scoured by the blasting of wind-blown sand and silt. Wind scour is shown by facets on stones and by polish, flutes, pits, and grooves eroded by flying grains of sand and silt. Don't mistake gas-bubble holes in lava for eolian pitting.

Inbound: Opposite the Artists Drive entrance.

● The black rock, known as Dinosaur Ridge across the valley at the base of the Panamints at 10–11 o'clock is Funeral basalt.

▶ | Note your odometer reading at Artists Drive entrance.

● Within the next mile look ahead at 1 o'clock to hills of black Funeral basalt at the base of the Black Mountains. A close inspection will reveal faint horizontal scars of lacustrine shorelines on their slopes. The nearly flat top of this area is also a lake-formed feature, called Manly Terrace (see Photo E–12). It has been a site favored by Native Americans.

Inbound: Look almost directly behind from West Side Road junction.

Coffin Cyn.
T Fan
Hanaupah Scarp
Badwater⌐
Badwater Turtleback⌐
Trail Cyn. Fan⌐
⌐Artists Drive Jct.
⌐Artists Drive Block
⌐ Manly Terrace
Zabriskie Pt.
T
Furnace Cr. Ranch
T

0   1   2 miles

Photo E–12.
High-altitude vertical photo of central Death Valley, north toward upper right corner, scale at lower right. (U.S. Air Force photo taken for U.S. Geological Survey, 374V-192.)

● About 2 miles north of Artists Drive entrance and a half mile short of West Side Road junction, excellent stratification can be seen in the rocks composing the Black Mountain front. The great variety of colors displayed (tan, cream, lavender, green, brown, gray, and white) suggest that these deposits contain much fragmental volcanic debris.

Inbound: West Side Road is 1.6 miles beyond Mushroom Rock, and this view is 0.5 mile farther.

● A little beyond Artists Drive exit, you round the point of Manly Terrace, pass Mushroom Rock, a weathered chunk of lava, and about 1.5 to 2 miles from Artists Drive exit get a good view ahead on the right of the Furnace Creek Formation that here makes up the Black Mountains front. The Furnace Creek is a Pliocene (3 to 5 m.y.) unit, consisting mostly of 5,000 feet of light-colored, soft, silty lake beds with intercalated fanglomerate and volcanic layers. These beds are beautifully exposed at Zabriskie Point off the Furnace Creek Wash Road. They contain borate minerals, and the hills are pocked with old abandoned borate mines and prospects.

Inbound: Mushroom Rock is 0.6 mile south of MM 61 and the Furnace Creek beds are seen well south from Gower Gulch.

⑤ About 6 miles from Artists Drive entrance you come to a dip sign and a 100 foot section of concrete pavement. The fan surface here is scoured, dotted with large boulders, and locally covered with polygonally cracked dry mud, all suggestive of flooding. Right of the highway, a vertical-walled gulch cut into the fan deepens to more than 20 feet near the mountains. This fan is fed by Gower Gulch, which heads at the Zabriskie Point overlook. Flooding and dissection are the result of artificial diversion of Furnace Creek drainage into the head of Gower Gulch. This was done to protect buildings in lower Furnace Creek Wash and on its fan. However, it would have been only a matter of time until the diversion would

have occurred naturally, for headward working Gower Gulch was on the verge of capturing Furnace Creek. Stream capture is a common geological phenomenon. It produces anomalous drainage patterns and effects changes in the regime of both the captured and capturing stream. The dead mesquite trees west (left) of the highway beyond here were deprived of water by this diversion.

Inbound: Gower Gulch's concrete road section is 2.5 miles from the 190 junction.

● Roughly 6.6 miles from Artists Drive entrance is Golden Canyon, an interesting little drive up a steep-walled gulch cut in Furnace Creek beds.

Inbound: Well marked for southbound travelers.

⑥ At 7.5 miles from Artists Drive entrance, and extending north for 0.4 mile, you begin to pass a beautiful little fault scarp, 2 to 7 feet high, breaking the fan surface 20 to 100 feet right of the road. In spots the gravel in the face of the scarp is different from that on the fan surfaces because the fault displacement has brought up gray gravel with well-rounded stones from the larger Furnace Creek fan that here underlie a thin deposit of brownish debris derived from the near hills.

Inbound: The B-2 wood stake 0.6 mile from the 190 junction is opposite the higher north part of this scarp, which extends south 0.3 mile to near the yellow/black road curve sign.

● Approaching Furnace Creek Inn, the steeply tilted, greenish beds are associated with fanglomerates in the Furnace Creek Formation. You turn left on Highway 190 and proceed down the surface of Furnace Creek fan 1.5 miles to Furnace Creek Ranch (see Photo E–12) and the Monument Visitor's Center, 0.1 mile beyond. Both are worthy of your attention, especially the center.

Inbound: Note your odometer at turnoff from Highway 190 onto Badwater Road.

LOS ANGELES BASIN TO MAMMOTH

Guide segments are arranged so that travelers bound for Mammoth can depart the southern California metropolitan areas by three routes: ① Devore, via Interstate 15/215 and Highway 395 (Segments A, G, and H), ② La Cañada, via Angeles Crest and Angeles Forest Highways (Segment I), and ③ San Fernando Pass (Segment J), via Antelope Valley Freeway (Highway 14). The San Fernando Pass and La Cañada routes join at Vincent Junction near Palmdale. They in turn rendezvous with the Devore/395 route in Indian Wells Valley. Travelers using this last exit follow the initial part of the Death Valley trip (Segment A) as far as the 395 separation. From Indian Wells Valley the consolidated routes proceed up Owens Valley through Bishop to Mammoth (Segments M through P).

Devore Exit

Use the introductory section and first part of Segment A of the Death Valley trip as far as the Highway 395 separation, then proceed with Segments G and H.

Segment G—Via Highway 395 to Red Mountain, 68 Miles

● This route (Figure 3–11) crosses the eastern part of the western Mojave Desert, where topography is subdued and rock exposures scattered.

> ▶ Record odometer reading as you turn onto Highway 395.

● In 1.5 miles you descend into and climb out of a deep, flat-floored gully—actually it's an arroyo, known as Oro Grande Wash. This is one of a large family of similar gullies cut into the *alluvial apron* sloping north from Cajon Summit. The cutting had to occur while streams from the San Gabriel Mountains could still flow north into the desert, and that had to be before the excavation of the Cajon amphitheater that beheaded the alluvial apron. Tectonic uplift or climatic change can cause dissection of alluvial aprons. Repeated surveys of bench marks along railroad lines through Cajon Pass indicate that this area is currently rising at rates between 0.5 and 2.5 feet per century, indicating that tectonic uplift is the likely cause of the dissection. Joshua trees plus scattered junipers and low brush predominate here.

Inbound: Oro Grande wash is 1.5 miles beyond the California Aqueduct.

● In another 1.5 miles you cross the east branch of the California Aqueduct (officially the Edmund G. Brown Aqueduct) that carries water to the Cedar Springs reservoir at the north base of San Bernardino Mountains.

Inbound: The aqueduct is 4.5 miles south of Highway 18 crossed at Doby Corners.

Looking back to 7–8 o'clock from beyond the aqueduct gives a good view of the San Gabriel Mountains, especially striking when they are snowcapped in winter. Eastward at 3 o'clock, the low saddle on the eastern skyline marks the Victorville-Twentynine Palms trough (see Mojave Desert province), and at 4 o'clock are the San Bernardino Mountains.

Inbound: The San Gabriels and San Bernardinos are in full view for a long distance southbound. Twentynine Palms trough-saddle is on the skyline at 9 o'clock.

● Beyond the stoplight at Highway 18, if visibility is at all passable, at 2:45–3 o'clock you get a view of the Victorville area with its hills

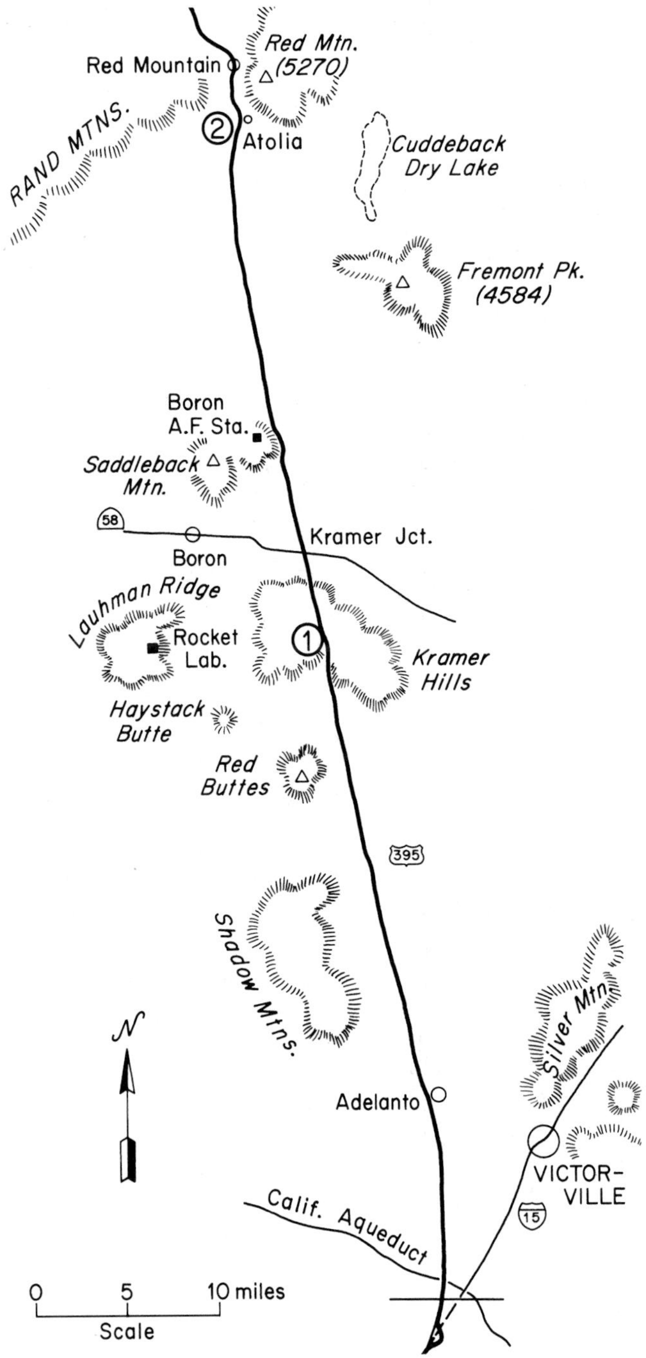

Red Mountain ○

Red Mtn. △ *(5270)*

RAND MTNS.

② ○ Atolia

Cuddeback Dry Lake

Fremont Pk. (4584) △

Boron A.F. Sta. ■

Saddleback Mtn. △

㊹58 ○

Boron

Kramer Jct.

Lauhman Ridge

■ Rocket Lab.

① *Kramer Hills*

Haystack Butte

Red Buttes △

㊋395

Shadow Mtns.

Silver Mtn.

Adelanto ○

◯ **VICTOR- VILLE** ㊋15

Calif. Aqueduct

N ↑

0   5   10 miles
Scale

Figure 3–11.
Segment G, I-15/395 separation to Red Mountain (town).

of granitic and metamorphic rocks. The light spots are mostly marble quarries supplying carbonate-rock to the cement plants at Oro Grande and Victorville. Year after year cement is second only to sand and gravel as the most valuable nonfuel (oil and gas) natural resource produced in California; gold is fourth. The high peak at 1:15 o'clock with a carbonate-rock quarry on its north side is Quartzite Mountain.

Inbound: This view is at 9 o'clock crossing Mojave Road, about three miles south of Air Base Road in Adelanto.

● Two to 3 miles beyond Adelanto the hills and ridges of the Shadow Mountains are on the skyline at 9–11 o'clock. They are composed mostly of granitic rocks that intrude a large metamorphosed sedimentary sequence of complex structure and many rock types, including marble, schist, quartzite, and a varied series of metavolcanics. About 4.5 miles north of Adelanto, spurs and knobs of these metamorphic rocks come close to our highway. The light colored isolated hill at the north end of the Shadows, seen at 9:15 o'clock at Shadow Mountains Road 10.5 miles from Adelanto, is Silver Peak (4,043), composed of white marble.

Inbound: Silver Peak is in view at 3 o'clock 1.5 miles south of Sun Hill Ranch Airport Road. The Shadow Mountains are on the right now for 10 miles.

● Beyond Sun Hill Ranch Airport Road at 10:30, or at 9 o'clock 4 miles farther north, are the two Red Buttes, composed of dark Miocene volcanic rocks. They are part of a string of smaller buttes extending southwest from the Kramer Hills, which are another 3.5 miles ahead.

Inbound: Red Buttes are at 3 o'clock, 3.5 miles south of Kramer Road that is at the south edge of Kramer Hills.

● That part of Kramer Hills close along the highway is composed largely of mid-Tertiary (10 to 20 m.y.) sedimentary and volcanic rocks of the Tropico Group, named from Tropico Mine about 40 miles to the west. These rocks rest upon more extensive older granitics and are locally capped by fanglomerates. The first knob, just left of the highway, exposes white, well-layered, silica-rich lake beds overlain to the south by darker lavas, both gently dipping. The lake beds can be seen close at hand in a road cut on the right. These rocks are representative of the Tropico Group in this area, and are seen again in another mile at mid-distance right of the highway. In 1.7 miles you start passing through some large road cuts. The first exposes dark coarse fanglomerates with volcanic fragments, the next has dark mixed up volcanics including fractured lavas, and the third features a different type of bouldery fanglomerate.

Inbound: You enter Kramer Hills nearly 4 miles south of Kramer Junction. Read about the road-cut exposures in reverse order.

● Descending from Kramer Hills toward Kramer Junction (Four Corners) you get splendid views out over the Mojave Desert. Of particular interest is the solar power-generating facility just northwest of Kramer Junction. Depending on the time of day, you may see a sea of mirrors that look like a rectangular lake (Photo G–1), but at any time you will see row upon row of something (mirrors). Columns of steam (at least four) rise from cooling structures where steam is recondensed to water for further use. The mirrors are arranged in north-south rows and rotate on horizontal axes. From Kramer Hills they are most impressive in the middle of the day, when the mirrors are nearly horizontal. They are curved so as to reflect sunshine onto a horizontal double-walled, vacuum-insulated, oil-filled glass pipe that runs in front of each line of mirrors. The oil is run past so many mirrors it is heated to several

Photo G–1.
Bright area is a sea of mirrors, part of a solar-power generating operation alongside Highway 395 just northwest of Kramer Junction (Four Corners).

hundred degrees Fahrenheit. This hot oil then passes through a heat exchanger, where it brings water under pressure to well above normal boiling point. That superheated water is then flashed to steam, which turns the turbines that generate electrical power. There are three such plants in this part of the Mojave Desert. As of 1992 this operation claimed to be the largest electrical solar power-generating establishment in existence, making 150 megawatts of power. The installations take up space, but this is clearly a renewable, nonpolluting source of energy. Southern California Edison Company buys the power under contract.

① Shortly you come to Kramer Junction (Four Corners).

▶ | Note your odometer reading here.

About 6 miles west is the settlement of Boron, site of one of the world's greatest borate-mineral mines. (See the resources section, Mojave Desert province.) Borates were discovered in the area during drilling of a water well in 1913. Further exploratory drilling in 1925 by

Pacific Coast Borax Company revealed larger and more desirable borate deposits at depths of less than 400 feet, and in 1926 sinking of a shaft led to development of the area into what has been the world's most productive borate deposit. In 1957 a huge open pit was developed in the shallow deposits at the west end of the district. The keys to the Boron operation is not only the shallowness and richness of the deposits but also the fact that much of the boron is in sodium borate compounds, rather than the more usual calcium borates. Sodium borates are more easily and cheaply refined and marketed. The borates occur as layers, lenses, nodules, and veinlets within lake-clay beds (shales) within the upper Tropico Group (10 to 15 m.y.). The operation is now owned by U.S. Borax and Chemical Company.

Inbound: Unfortunately, the huge open pit at the Boron mine can be visited only by special arrangement.

● The sea of mirrors begins about 0.8 mile beyond Kramer Junction and extends north for 0.4 mile. You can see the setup best by turning left (west) off 395 into the wide parking area at the northeast corner of the field of mirrors (see Photo G–1).

The dark, isolated hill about 3 miles off to the left at 10:15 o'clock beyond the mirrors is Saddleback Mountain, obviously named for its shape. It should not be confused with Saddleback Butte 30 miles to the south-southwest. It is largely a granitic knob capped by younger lavas that shed a mantle of dark debris down its slopes.

Inbound: Saddleback Mountain is at 3 o'clock just a little beyond the Camp Boron Federal Prison, heralded by a rare curve in the highway.

● About six miles from Kramer Junction is the Camp Boron Federal Prison, accompanied by a tracking dome on top of a nearby granitic ridge.

Inbound: Location of the Federal Prison is well advertised by the highly visible white tracking dome.

● For the next 15 miles you traverse a relatively monotonous terrain of plains and low rounded knobs. The surface material is largely disintegrated granite, but solid granitic bedrock lies not far beneath over much of this area. Coarse grained homogeneous granites disintegrate so readily in a desert environment that they come to underlie relatively featureless terrain.

Inbound: If you tried to dig a hole with a shovel in much of this country, you would be surprised how quickly you struck granitic bedrock.

● At 8 to 10 miles from Kramer Junction on the skyline at 2:15 to 2:30 o'clock is Fremont Peak (4,584) composed of old (Precambrian) granitic rocks intruded by much younger (Mesozoic) granites.

Inbound: Seen at 9 o'clock about 6 miles south from 20 Mule Team Parkway. Its appearance has changed from the sharp peak seen at Red Mountain.

● About 17 miles out of Kramer Junction, near Call Box 652, the Rand Mountains (granite and schist) make up the skyline from 10–12 o'clock, and Red Mountain (5,261), a volcanic knob, is prominent at 1 o'clock.

Inbound: Both were already seen and are now behind you.

● Beyond "*Ridgecrest 27, Lone Pine 98, Bishop 158,*" low artificial mounds of gravel on the hill slopes at 11–12 o'clock are the result of placer mining for tungsten. Within the next few miles larger piles of placer gravel are seen on both sides of the highway approaching Atolia. These workings are in alluvial gravels in which pieces of a heavy, tough, resistant tungsten mineral, scheelite ($CaWO_4$), have accumulated. The Atolia district has been a major tungsten producer.

Inbound: The gravel piles of the placer workings south of Atolia are obvious left of the highway.

● As you pass through the town of Atolia, at about 22 miles from Kramer Junction, look right at 3:15 o'clock to see Cuddeback Dry Lake about 7 miles away.

Inbound: Look at 9:15 o'clock for Cuddeback. The sharp symmetrical mountain far off at 10:45–10:30 o'clock between Red Mountain and Atolia is Fremont Peak.

● Within two miles, approaching the town of Red Mountain, you see to the left hoists, head frames, tanks, spoils piles, and buildings of abandoned mines (Photo G–2). Red Mountain and Atolia, along with Johannesburg and Randsburg just ahead, make up one of California's more noted mining districts. The area started as a gold camp in 1895, and gold was the chief product up to World War I. The war stimulated production of tungsten in the Atolia area, and after the war artificially high prices excited interest in silver, resulting in discovery of rich bonanza deposits near Red Mountain. The value of metals produced between 1895 and 1924 exceeded $35 million, and total production from the district may have approached $50 million, about equally divided among gold, silver, and tungsten. Gold and tungsten production continued intermittently up to the mid-1950s, in response to temporarily high prices, especially for tungsten during World War II and the Korean conflict.

Photo G–2.
Remains of a mine hoist, waste pile, and ore processing plant near Red Mountain in the Randsburg District.

The gold and silver deposits were shallow, the deepest workings being 600 feet, and some were fabulously rich. The gold is associated mostly with quartz veins in granitic rocks and the Rand Schist, a metamorphic formation that you will see shortly. The tungsten occurs as veins in younger (Mesozoic) granitic rock around Atolia and Red Mountain and as placer deposits in alluvial gravels. The silver is in veins in the Rand Schist near Red Mountain. Although occurring in old rocks, the ore deposits are relatively young, having formed during an episode of Miocene (5 to 23 m.y.) surface and near-surface igneous (volcanic) activity.

Inbound: If you enjoy antiquity and old mining camps, Red Mountain, Johannesburg, Randsburg and their surroundings are great places for scouting around.

Segment H—Red Mountain to Junction with Highway 14, 30 Miles

▶ | Record your odometer reading in Red Mountain |

● Within two miles you pass through Johannesburg; Randsburg is a short distance west (Figure 3–12). Between Red Mountain and Johannesburg, piles of coarse, multicolored rock debris on the hillsides are mostly waste dumps from mines. Piles of fine material are tailings (ground up rocks) from mills that processed the ore.

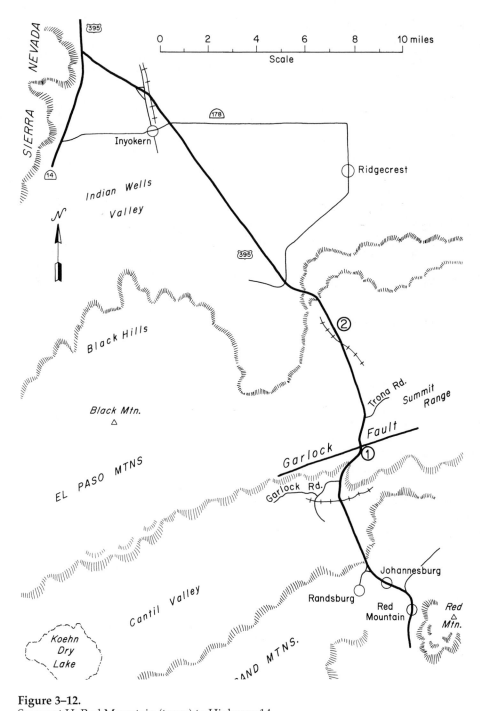

Figure 3–12.
Segment H, Red Mountain (town) to Highway 14.

Inbound: Many waste dumps and tailings piles are seen on hillsides, and even in towns, between the Randsburg turnoff and Red Mountain town.

● Just 1.2 miles beyond Johannesburg you pass the Randsburg turnoff, to the left, and soon the highway straightens out on a north-northwesterly course downhill. El Paso Range lies across the valley at 9–11 o'clock, and Summit Range is at 11-12:30 o'clock. If the weather is clear, you may see just a little of the southern Sierra Nevada crest on the far skyline at 11 o'clock, possibly snow capped. In about a mile the view westward at 9 o'clock is into Cantil Valley as far as Koehn Dry Lake.

Inbound: Unfortunately, all this is behind you.

● The slopes on either side of the upper part of this straight stretch are partly mantled with slabs of Rand schist; you occasionally catch flashes of light reflected from these smooth fragments. White spots are the outcroppings of quartz veins and fragments derived therefrom. The first large road cut on the left, just a bit more than a mile from the Randsburg turnoff near the start of a section of four-lane highway, exposes a cross section through the schist.

Inbound: The Rand Mountains loom across the valley after you cross the railroad, and the large road cut in Rand Schist is just beyond "Elevation 3,000 Ft." Hillsides are littered with schist slabs. Old prospects and mine workings scar the hillsides.

● At the bottom you cross railroad tracks, the highway curves a bit to the right, and in less than a mile the Garlock road (paved) joins from the left. Upgrade from this intersection, the dark rocks capping ridges on both sides and shedding boulders down-slope are lavas.

Inbound: Dark lavas on both sides are easily recognized approaching the Garlock Road turnoff.

① As you continue uphill a 40 foot linear scarp, dappled in gray and white, approaches the highway obliquely from the left at 9–12 o'clock. The scarp marks the trace of the Garlock fault (Photo H–1), a major left-lateral fracture (see Mojave Desert province). Near the top of the ascent, 1.5 miles from Garlock Road and 8 miles from Red Mountain, where the highway curves right and then left, alongside a wide turnout area on the right, you intersect the fault. Fanglomerates exposed in the road cuts on both sides are chewed up, and at times cracks have developed in newly surfaced paving here, suggesting possible slow creep along the Garlock fault in this area, although this has not been proven.

Inbound: You cross the Garlock fault at the top of the descent, 1.6 miles beyond Searles Station Road, where a four-lane highway section starts.

● The Searles Station Road to Trona takes off right in 1.6 miles. A distant view of Telescope Peak in the Panamint Range, often snow capped in winter, can frequently be seen at about 2 o'clock here and on down the highway.

Inbound: Highway 395 straightens into a 4.5 mile run about 1.5 miles past China Lake Boulevard, along which Telescope Peak can be in view at 7:30 and lower clock hours.

In another 3.3 miles you cross an abandoned railroad right-of-way. This was part of the old rail line up Owens Valley. The old road bed still has rails in many places, and its course is obvious on the hillside to the left as you continue north, for example opposite Call Box 133. About 2 miles beyond that spot the rocks on both sides of the road, mostly granitics, are more variegated in color and general appearance. This reflects secondary alteration of a type commonly associated with ore deposits. The hills ahead and to the right, where the straight run ends with a distinct curve to the

U.S. Hwy. 395
Slate Range
Avawatz Mtns.
Garlock Fault
So. End Death Valley
Searles Lake
Eastern
Mojave
Desert
Ridgecrest

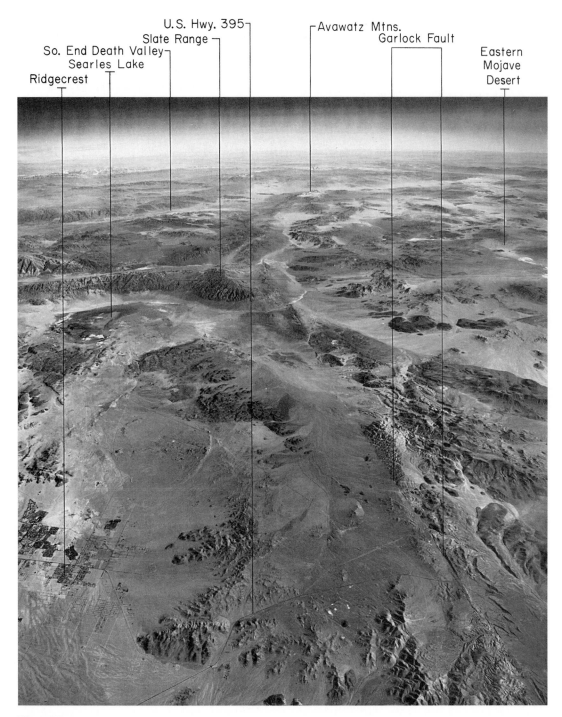

Photo H–1.
High-altitude oblique view east along Garlock fault. (U.S. Air Force photo taken for U.S. Geological Survey, 744L-026.)

left, house a number of prospects and several named mines, the latter being out of sight of this highway.

Inbound: The mineralized area is to the left approaching and beyond the turnoff to Ridgecrest at China Lake Boulevard. The railroad right-of-way is obvious on the hillside to the right from there to its crossing half way along the straight run toward Searles Station Road.

● Shortly you come to the turnoff to Ridge-crest (see Photo H–1) on China Lake Boulevard and start a long, gentle descent on a straight run.

▶ | Note your odometer reading at this intersection.

You look down into Indian Wells Valley, and the southern Sierra Nevada fill the skyline from 9–12 o'clock; Owens Peak is at 11 o'clock, Olancha Peak is at 12:30 o'clock, and the Argus Range (largely granitic rocks) composes the skyline at 2–3 o'clock. Within two or three miles, look at 9 o'clock to see lava flows capping Black Mountain.

Inbound: If you stopped and looked back before crossing the railroad overpass, you have seen these features; otherwise, they are behind you.

● In about five to six miles you see the southern Sierras more plainly and can recognize that the light-colored, clean, hard granitic rocks of their core have a different topographic expression than the more weathered granitic and metamorphic rocks composing the range front. The line of the fairly recent (1970) Los Angeles Aqueduct well up on the face of the Sierras is plainly visible at 11 o'clock.

Inbound: You see these features best from the inbound start of this segment at the 395/14 intersection.

● About eight miles down the straightaway from China Lake Boulevard you pass the junction with Highway 178 and in less than two miles beyond rise onto an overpass across the abandoned railroad. From here you get a good view of the gap at Little Lake (at 1:30 o'clock) into which you shortly head. It separates the Sierras from the Coso Range. Volcanics mantling the south flank of the Coso Range are seen at 2 o'clock.

Inbound: You will have to stop before rising onto the railroad overpass, get out and look back to get this view.

● You proceed to a junction with Highway 14. Now shift to Segment M to continue the trip north to Mammoth.

LA CAÑADA EXIT

Segment I—La Cañada to Vincent Junction, 33 Miles

● This segment follows a winding highway across San Gabriel Mountains. Alertness is required to spot some locations, and the geology is complex. Travel leisurely, make as many of the stops as possible, read ahead in the guide, and navigate by keeping track of check points such as bridges, intersections, passing lanes, power lines, roadside call boxes, mileage markers (MM), and distances. Some people, including me, get car sick reading on winding roads. You may prefer to read parked in a turnout. Things get simpler after you cross Big Tujunga Creek. Special locations are numbered on the map (Figure 3–13) and correspondingly in the text. Study the map and scan the text before you start this journey to get a feeling of scale and distribution of features.

● In La Cañada, turn north off Freeway 210 onto Highway 2 (Angeles Crest).

▶ | Note your odometer reading.

You are headed due north up the alluvial slope at the foot of San Gabriel Mountains. Near its head, this straight stretch is reputed to have the steepest grade of any part of the Angeles Crest-Angeles Forest Highway. Read ahead in the road guide.

Inbound: Watch your speed. This stretch is steeper than it appears, and the gendarmes patrol it.

● In 0.7 mile the highway curves east, you cross La Cañada Arch Canyon on a concrete bridge, and immediately you enter road cuts in crystalline rocks. In passing from the alluvial slope to the mountain block, you crossed the Foothill fault zone that determines the south

face of the mountains. Displacements up to 6 feet occurred on fractures in this zone near San Fernando in the 1971 earthquake.

Inbound: Digest this information between Starlight Crest Drive and La Cañada Arch Canyon.

● The igneous rock exposed in road cuts opposite La Cañada Country Club turnoff (Starlight Crest Drive) and beyond is part of a rock unit named Wilson diorite, after Mt. Wilson with its TV towers and telescopes. The Wilson diorite is a member of a relatively young group of igneous intrusive bodies in the San Gabriels, about 80 to 90 m.y. old. Beyond the turnoff, you will see dikes of white igneous rock cutting the diorite.

Inbound: Location is no problem. This is your introduction to the Wilson diorite.

● Road cuts just beyond the upper end of the golf course are poor, but within a mile are good exposures of Wilson Diorite.

Inbound: White dike rock intruding the Wilson Diorite may catch your eye in road cuts to the right approaching La Cañada Country Club turn in (Starlight Crest Drive).

① About 2.5 miles from La Cañada, still short of the Angeles Crest forest station, is the first of two large macadam turnouts east of the highway. The first is the best for a stop, but the second is nearly as good. From them you get a good view into the narrow canyon of Arroyo Seco, a perennial stream and part of Pasadena's water supply. Downstream and directly across the canyon are gently sloping topographic benches high on the canyon walls. These are remnants of former wide valleys formed during pauses in the cutting of the Arroyo Seco. These pauses were terminated when renewed uplifting of the mountains along the Foothill fault zone

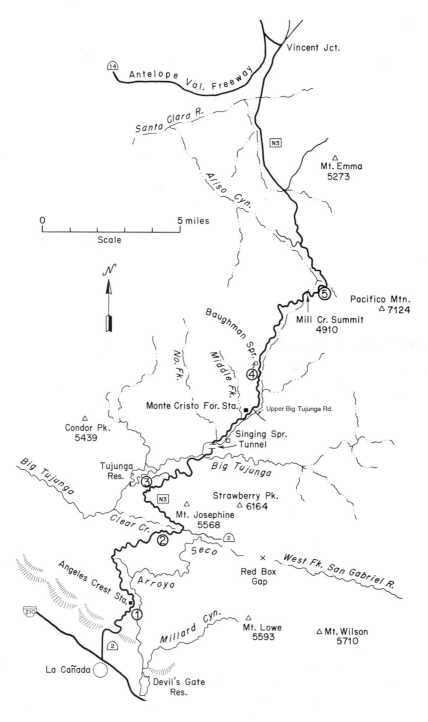

Figure 3–13.
Segment I, La Cañada to Vincent Junction.

caused the stream to cut down again. On clear days views over Pasadena, downtown Los Angeles, and possibly even to Catalina can be great.

Inbound: The upper turnout is only 0.3 miles below the Angeles Crest station and the lower turnout is about 0.1 mile farther, on the left.

● Between the two large macadam turnouts you enter a mixture of old metamorphic gneisses and igneous rocks intruded by a complex of pale dikes and small irregular bodies of younger igneous rocks. This unit is the San Gabriel Formation. You pass through it for nearly five miles. All crystalline rocks in San Gabriel Mountains have been severely deformed, badly fractured, and even locally shattered. Don't be disturbed if they look mixed up to you; even to professional geologists they are confusing.

● You shortly pass Angeles Crest forest station.

▶	Record your odometer reading here.

The mixed light, dark, and pink rocks seen in road cuts beyond are representative of the San Gabriel Formation. At the first concrete bridge, in about a mile, pink dikes intruding dark rock should catch your attention. The dark rock is relatively rich in minerals containing considerable iron and magnesium, and the younger pink rock is relatively richer in minerals containing potassium, aluminum, and silicon. There are lots of turnouts along here; one, at Mileage Marker 28.36, about 0.7 mile from the forest station, provides a good view down to the winding floor of the Arroyo and up into its headwaters.

Inbound: Angeles Crest forest station is 6 miles from the Angeles Forest-Angeles Crest intersection. The above features are seen within the last mile approaching it.

● Within the next two miles the highway swings westward back into several larger canyons, crosses two more concrete bridges,

and then bears eastward out onto a narrow ridge. Rocks in road cuts are badly fractured and complexly riven by intrusive dikes. Those canyons harbor trees including oaks, big cone spruce, bay, big leaf maple, and long needle Coulter pines (planted). The chaparral on surrounding slopes was badly burned in 1979 and by 1992 had almost completely recovered. Tall dead trees were big cone spruce. Just beyond MM 30.65, four miles from the forest station, is a deep, double-walled road cut exposing well banded gneiss cut by dikes.

Inbound: These forested canyons with concrete bridges are better seen inbound than out. This is also true of the tall big cone spruce snags killed by the Arroyo Seco fire.

● About 1 mile beyond the passing lane, under a power line crossing that starts at MM 31.12, you enter a transition zone to a rock unit of more uniform character, distinguished by lighter color and rugged craggy slopes, as seen particularly on hillsides to the right. You have crossed several branches of the Sierra Madre fault and are now entering a different body of igneous rock, the Lowe granodiorite, about 225 m.y. old.

Inbound: Exposures of pale Lowe granodiorite extend for about 0.7 mile before you are back into darker San Gabriel Formation.

② About 8 miles from La Cañada, 5 miles from the forest station, is the paved Angeles Forest Clear Creek scenic overlook. This is a good place to stop. Below to the north is the canyon of Clear Creek with Mt. Josephine on the skyline. The brown road cuts on the other side of Clear Creek are in the San Gabriel Formation. About 200 to 300 feet up-slope from them is pale rock making craggy outcrops extending to the skyline. This is the Mt. Josephine granodiorite, only 80 m.y. old, a youngster compared to the 1,700 m.y. age of parts of the San Gabriel Formation. The straight course of the contact between these two rock units, marked partly by a linear band of especially green vegetation, suggests that it is a fracture up which water seeps, a strand of the San Gabriel fault.

Photo I–1.
Looking east up headwaters of Arroyo Seco to Red Box gap from junction of Angeles Crest and Angeles Forest highways.

Inbound: A stop at the Clear Creek overlook 0.9 mile from the intersection is worthwhile.

● When you leave the overlook, road cuts again expose dark rocks cut by younger pink dikes, presumably the San Gabriel Formation. In one mile is the junction of Highway 2 (Angeles Crest) and County Highway N-3 (Angeles Forest Highway). You take the latter but first pull off into the wide macadam turnout and parking area on the right, opposite the intersection. The small building is a Forest Service information center.

On the skyline to the east is Red Box gap (Photo I–1) where the road to Mt. Wilson turns off. The straight canyon extending east toward Red Box is the uppermost part of the Arroyo Seco, which, up to this point, has followed an irregularly winding northward course. You are in a gap or saddle between the head of the Arroyo and Clear Creek, which has carved another unusually linear canyon extending west

into the Big Tujunga. It is aligned with the head of the Arroyo and the two gaps. If you were at Red Box, you could see that the same alignment extends eastward, first down the West Fork of San Gabriel River and then up its East Fork. This is no accident. All these features are along the trace of one of southern California's major faults, the San Gabriel (see Figure 1–1). Rapid erosion of ground-up rock along the fault zone has created this alignment of canyons and gaps. The many pine trees at the junction are coulters, part of a Forest Service plantation.

▶ | Record odometer reading at this junction. |

Inbound: A stop in the parking area at the intersection is as worthwhile inbound as outbound. Road cuts between here and the Clear Creek scenic overlook expose a lot of San Gabriel Formation metamorphic rocks. Record odometer at intersection.

● You now follow Angeles Forest Highway toward Palmdale and in 0.3 mile enter a section of four-lane highway. Mileage markers (MM) are now red figures on white. The first road cuts expose gravel succeeded by brown weathered rocks, part of the San Gabriel Formation. Some subsequent cuts are also in gravel.

● Within 1.3 miles from the junction the landscape is clearly into the lighter-colored Mt. Josephine granodiorite. At *"Icy"* and *"Slide Ahead"* start watching road cuts and slopes to the right to see masses of large angular slide blocks of granodiorite.

Inbound: The change from pale Mt. Josephine granodiorite to dark brown weathered San Gabriel Formation is easy to recognize in about another mile along toward the Angeles Forest/Angeles Crest intersection.

③ About 2.3 miles from the junction is a large turnout on the left just beyond *"Slide Area Next 2 Miles"* that gives a good view of the Big Tujunga Canyon and the Big Tujunga flood control dam area with its many roads. Walking out on the ridge a short way may provide a view of impounded water. To be effective, a flood-control reservoir should not be filled with water or rock debris. One of the problems with the Big Tujunga reservoir is debris. It lost nearly 25 percent of its capacity in a single year, 1938, to deposits of sand and gravel. A lot of debris has now been hauled out of the impoundment area and piled into a side canyon.

Inbound: This turnout, for a look into the Big Tujunga near the flood-control dam, is 1.6 miles beyond the Sunland-Big Tujunga road turnoff.

● In another 0.2 to 0.3 mile, road cuts expose old, angular, blocky, rockslide or rock-fall deposits (Photo I–2), which, as the signs suggest, readily shed debris onto the highway. A huge road cut in 0.6 mile exposes good slide debris.

Inbound: Much landslide debris of angular Mt. Josephine blocks is seen on slopes and in road cuts for the next 1.6 miles.

●

Photo I–2.
Rock-slide debris derived from Mt. Josephine granodiorite along Angeles Forest Highway above Big Tujunga flood-control dam.

You continue in the Mt. Josephine granodiorite for several miles. Locally it displays some darker phases, but some local dark spots are clearly inclusions of older gneissic rock shot through with dikes of Mt. Josephine granodiorite. A good example is 3.5 miles from the start of N-3 and extending for 0.3 mile.

Inbound: This location is near the turnoff of the Sunland-Big Tujunga paved road, 2.5 miles from Big Tujunga bridge; a good exposure lies just beyond the turnoff.

● At 6.3 miles from the start of Angeles Forest Highway you approach and cross the narrows of Big Tujunga Creek on a high-arch concrete bridge. The darker rock at both ends of the bridge is a phase of the Mt. Josephine intrusive body that is somewhat poorer in silica and richer in iron, calcium, and magnesium than usual. A stop at the turnout beyond the bridge and a walk back to the lookout over the gorge are worthwhile. This gives you a closer view of the dark phase of the Mt. Josephine granodiorite and reveals some smoothly shaped inclusions, 3 inches to 2 feet across, of still darker older rock. Big Tujunga gorge is deep and narrow and the riparian vegetation, mostly big cottonwood trees, is attractive, especially in the fall.

Inbound: A stop in parking area on the left at near end of Big Tujunga bridge and a walk to viewpoint are worthwhile. Across the bridge travelers look down into the deep narrow gorge.

● Entering the tunnel 0.3 mile up the highway, you see exposures of strongly banded gneiss at its south portal. The gneiss is a large chunk of old rock (1,700 m.y.) caught up in the Mt. Josephine granodiorite (80 m.y.). The north portal is in Mt. Josephine granodiorite. In 0.1 mile on the west (left) is the nice Hidden Spring picnic area with restrooms. Hidden Spring cafe is 0.4 mile farther along.

Inbound: Hidden Spring picnic area at tunnel entrance is easy to spot, and the view of the old gneiss in the right road cut emerging from the tunnel is good.

● As you pass Forest Springs watch carefully the road-cut exposures beyond. Much of the rock is granular, disintegrates easily, and is brownish to gray. Some is white and black, and more solid with a blocky fracture. The disintegrating and solid types are mixed, but by the time you pass Monte Cristo forest station in another 0.4 mile, only the white and black rock is seen. You have just crossed a zone in which Mt. Josephine granodiorite (disintegrating) intrudes a much older (1,220 m.y.) rock known as *anorthosite.* Samples from the moon suggest that anorthosite may be a major constituent of the lunar uplands; so geologists are much interested in this San Gabriel body. You travel through anorthosite and gabbro for the next 3.5 miles. Note the whiteness of the adjoining hill slopes.

Inbound: The change from anorthosite to Mt. Josephine granodiorite at and beyond Monte Cristo station is dramatic.

● Anorthosite is an unusual igneous rock made up almost entirely of a calcium-rich feldspar, which here forms crystals commonly several inches to occasionally 2 or 3 feet across. In places this body of anorthosite displays a steeply inclined banding of dark and light layers. The darker layers comprise a type of rock called *gabbro,* which contains some large (many inches across) crystals of a greenish mineral, *pyroxene,* and chunks of titanium-rich iron oxide mineral, *ilmenite.* The light gray to white layers are the anorthosite, some fresh samples have a delicate lavender color. This layering originated in an essentially horizontal position through settling of mineral crystals within a molten body. The solidified body with its layers were later tilted to their present inclination.

Inbound: Good anorthosite and gabbro here display the features described, especially banding, and they continue up to the Monte Cristo forest station, 2.3 miles ahead.

Anorthosite masses are rare, and you are fortunate to have one in our own backyard. Many lunar rocks are abnormally rich in titanium, and sure enough, associated with this anorthosite is a lot of ilmenite, a titanium-rich iron oxide.

④ You pass Monte Cristo forest service campground on the right and in 0.4 mile enter a four-lane section where the dark rocks are mostly gabbros, some spotted like a leopard. In 0.2 mile, where the highway curves west just short of Baughman Spring, the dried up watering spot with cottonwood tree and stone enclosure left of the highway, is a former wide turnout opposite MM's 14.65, 14.62, 14.39. It is now largely filled by waste-rock debris scraped from the highway. Parking is best on the west side, and access to the cut rising above the fill is easy via a haul road. This is a good place to see anorthosite close up. Banding is unusually good and the white pointed structures (Photo I–3), now largely buried, are

Photo I–3.
Tilted layers of anorthosite and gabbro near Baughman Spring between MM 14.39 and 14.65. White triangular features are crystallization structures pointing to top of original magma body.

accumulations of feldspar crystals that grew upward from the floor of a chamber. They indicate that the top of the anorthosite body is to the north. The south part of the exposure has good leopard-spot gabbro, enclosing some heavy, black, metallic crystals of ilmenite. In the little gully to the south you can find other minerals and rocks.

Inbound: The best place to see anorthosite is at the partly rock-waste filled turnout, 0.2 mile beyond Baughman Spring. Park by the highway; walk up the road on fill.

● In 0.8 mile beyond Baughman Spring, exposures of dark well-banded rock are followed by strongly banded, lighter materials. This is a transition zone marking the contact between the anorthosite body and a large mass of Lowe granodiorite through which you now drive to complete crossing the mountains. The rocks at the contact are banded because of a high degree of shearing to which they have been subjected. The banding in the Lowe granodiorite dies out northward up the road.

Inbound: After a long bend to left about 1.5 miles from Mill Creek summit, you come in another 1.5 miles to the highly laminated rocks of the contact zone and can view good anorthosite in another 0.4 mile.

The road cuts, on up to Mill Creek Summit (4,910) provide excellent exposures of the Lowe granodiorite. It is darker brown than the Mt. Josephine body and looks coarser grained. Note the typical craggy slopes left of the highway.

▶ | Record odometer reading at the summit.

Inbound: A large parking area at "Passing Lane Ahead" 7.7 miles from Vincent Junction gives a great view of dark dikes in Lowe granodiorite in the opposing road cut. The dikes in Photo I–4 are about at the 11.1 miles point. Record your odometer reading at Mill Creek summit.

Photo I–4.
Dark subhorizontal dikes intruding Lowe granodiorite, exposed in road cut north of Mill Creek Summit.

⑤ The Lowe granodiorite contains distinctive, dark, fine-grained subhorizontal dikes (see Photo I–4). These are most abundant in road cuts about 0.9 to 1.0 mile north of the summit.

● About 3 miles beyond the summit is Aliso Canyon road, and beyond that is good gneissic banding in the Lowe granodiorite.

Inbound: Gneissic banding in Lowe granodiorite is in the big double-walled road cut at 6.7 to 6.8 miles from Vincent Junction, approaching Aliso Canyon road.

● At 4.3 miles from Mill Creek Summit you emerge onto a 1.5 mile straightway, with a passing lane, descending a gently inclined alluvial slope. In about a mile Mt. Emma Road

turns right. Many of the road cuts beyond the straightaway expose dark-brown weathered alluvium through which occasional knobs of rock project.

Inbound: Dark alluvium is seen in road cuts before starting straightaway at 4.1 miles. The high distant skyline with trees is Mt. Gleason.

● The Edison Company distribution station sits on top of a westward sloping, well-graded, alluvial surface. The material in the road cut just beyond is deeply weathered crystalline rocks.

Inbound: Edison station is easily identified to the right when you clear the double-walled road cut in oxidized, weathered granitic rocks at 1.3 miles.

● Soon you emerge into the open gap at the head of the Santa Clara River drainage through which the railroad and freeway pass to the desert. This is Soledad Pass, and its size and openness suggest that earlier considerable drainage may have passed this way, perhaps from the desert to the sea, or from the San Gabriels to the desert. Currently, the headwaters of the Santa Clara River are eating back into the gap and eventually may work through to the desert. After a wet winter, wildflower displays are usually good here.

Inbound: Soledad Pass is easily observed and identified inbound at the start of this segment.

● At 33.4 miles from La Cañada you come to Vincent Junction and, with a zig and a zag, join Antelope Valley Freeway (Highway 14). If you are continuing to Mammoth, skip to Segment K; if returning to Los Angeles, follow Segment J in reverse, from end to beginning.

Inbound: Note your odometer reading at Vincent Junction as you turn onto Angeles Forest Highway.

Photo J–1.
"Bucket of Worms" freeway interchange of Interstate Highways 5 and 14 at east entrance to San Fernando Pass. A major span of this freeway (then under construction) collapsed in the 1971 San Fernando earthquake. (Photo by Helen Z. Knudsen.)

SAN FERNANDO EXIT

Segment J—San Fernando Pass to Vincent Junction (Palmdale), 29 Miles

● This exit from Los Angeles Basin for the Mammoth trip starts at the huge "bucket of worms" interchange (Photo J–1) at the entrance to San Fernando Pass. There Highway 14 (Antelope Valley Freeway) separates from Interstate 5 (Golden State Freeway). This point can be reached via the 5, 405, and 210 Freeways (Figure 3–14). You follow Antelope Valley Freeway (Highway 14) to a junction with Segment I at Vincent Junction near Palmdale.

Inbound: From the "bucket of worms" freeway interchange you are on your own with multiple choices.

San Fernando Pass is an area of considerable structural complexity because the Foothill fault zone, which bounds the south base of San Gabriel Mountains to the east, comes head to head with the Santa Susana thrust fault that has curved northward from its position about halfway up the south face of Santa Susana Mountains to the west. The situation is further complicated by several small, cross-cutting faults extending northeasterly. It is impossible to tell whether the Foothill fault truncates the Santa Susana or vice versa, or whether the faults grade into each other.

Both lateral and vertical displacements of as much as six feet occurred on fractures within the complex Foothill fault zone a few miles east during the San Fernando earthquake of February 9, 1971. The earthquake caused extensive damage to the incomplete bucket of worms interchange.

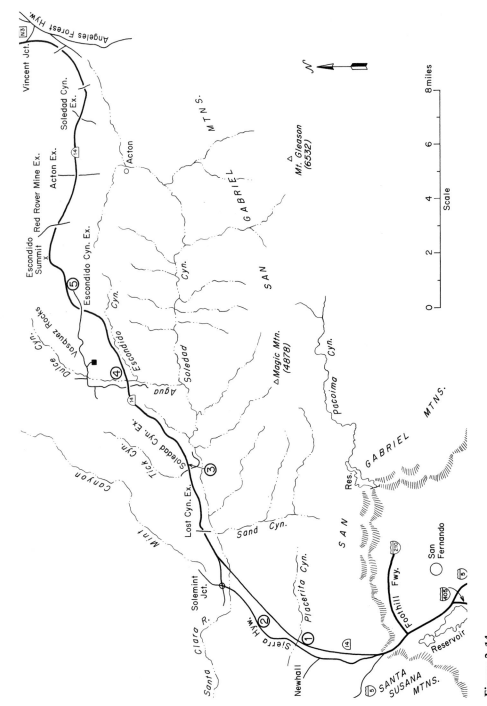

Figure 3–14.
Segment J, San Fernando Pass to Vincent Junction.

● Rocks exposed in the huge freeway cuts of the interchange are marine sedimentary beds of Plio-Miocene age (2 to 10 m.y.). There is much fine sandstone, siltstone, some shale, and considerable conglomerate. Some of the exposures show good examples of huge scour channels, filled with coarse debris, which cut deeply into underlying beds. The tilt of these beds differs greatly in direction and steepness from place to place, reflecting their complex structure.

Inbound: Watch the big cuts approaching, at, and beyond the I-5/14 interchange to get a sense of the complex structures here. The nature and inclination of the beds change from cut to cut.

▶ As you turn off on Antelope Valley Freeway toward Palmdale, note your odometer reading.

● The first large freeway cut on the left, where the northbound travelers from I-5 come onto Highway 14, displays south tilted Plio-Miocene (4 to 10 m.y.) sandstone, shale, and conglomerate beds with small faults, small and large sand-filled scour channels, and local cross-bedding. The walls of large scour channels can be mistaken for faults unless you view the whole cut.

Road cuts for a little more than two miles beyond the interchange, are in early Pliocene to late Miocene (5 to 8 m.y.) marine sedimentary beds similar to those at the interchange.

Inbound: These cuts are seen beyond San Fernando Road exit in the last mile approaching the intersection with I-5.

In a little over two miles from the interchange you pass the Newhall turnoff, cross Whitney Canyon, and enter road cuts in nearly horizontal, brown, coarse, nonmarine conglomerate and sandstone beds of the Saugus Formation, Plio-Pleistocene in age (1 to 3 m.y.). Cuts are particularly spectacular on the right coming down into Placerita Canyon.

Inbound: The road cuts show you what the Saugus Formation looks like, and roadside signs tell you where you are.

① In another mile you cross Placerita Canyon, so named for the discovery of placer gold in 1842 by the Spanish, six years before the find at Sutter's mill that touched off the great Gold Rush of 1849.

Inbound: The signs identify Placerita Canyon and Road for you.

● Beyond Placerita Canyon are more of the nearly flat conglomeratic Saugus beds. Here you are passing the partly abandoned tanks and wells of the almost depleted Placerita oil field, a forlorn sight and depressing testimony to improper exploitation of a natural resource. This field, discovered in 1948, proved to be a prolific producer from relatively shallow depths. However, the area had once been subdivided by a land promoter, and town-lot plats were held by many different owners. Everyone got into the act. Too many wells were drilled in the rush to get the oil out before somebody else got it, with the result that only a part of the full potential of the field was realized.

Inbound: Placerita oil field is closer on the right than outbound but is harder to see. Look back approaching Placerita Road.

② Approaching the top of the rise beyond Placerita Canyon the heretofore nearly horizontal Saugus beds abruptly steepen, presumably because of proximity to the San Gabriel fault (see Figure 1–1), a major structure with many miles of right-lateral displacement (see Segment I) that crosses the freeway near here. The Placerita oil field owes its existence in part to the trapping of oil in beds turned up along the fault. At the Canyon Country exit note the essentially vertical bedding. These strata are probably on the far side of the fault.

Inbound: Steeply tilted beds beyond Via Princessa off ramp have probably been deformed by the San Gabriel fault.

Photo J–2. Angular unconformity between gently tilted Mint Canyon Formation and horizontal Pleistocene gravels in road cut alongside Highway 14 at Soledad Canyon Road junction.

● Beyond Canyon Country exit, the wide upper Santa Clara River Valley opens out ahead. You shortly dip down and cross the sandy (usually dry) bed of the river on a long bridge.

Inbound: You cross Santa Clara River bridge about one mile beyond Sand Canyon Road.

● In a huge new road cut left of the freeway, just beyond Sand Canyon Road overpass, gently inclined, pale sandstone, conglomerate and thin shale beds of the terrestrial Upper Miocene (6 to 10 m.y.) Mint Canyon Formation are beautifully displayed. Ledges on hillsides left of the freeway in the next two miles reflect the gentle inclination of resistant beds in this formation.

You are now in the area of a local depositional site on the northwest flank of San Gabriel Mountains, known to geologists as Soledad Basin. Some 25,000 feet of land-laid deposits accumulated here, part of them remarkably coarse. You pass through many thousand feet of inclined Mint Canyon beds before getting to a still older bouldery deposit. With minor exceptions, the road cuts ahead for the next 6.5 miles are in Mint Canyon beds.

Inbound: Between Soledad Canyon and Sand Canyon roads you see a lot of the Mint Canyon Formation.

● Approaching the Soledad Canyon exit, a major sand and gravel operation occupies the bed of Santa Clara River just east of the freeway. In 1990 the total value of sand and gravel marketed in California was nearly $669 million. It was second only to oil and gas as a natural-resource product of the state.

③ Just beyond the Soledad Canyon overpass is a large cut on the left side with a striking angular unconformity (Photo J–2) between nearly horizontal, iron-stained, Pleistocene (1 m.y.) gravels resting on pale, gently tilted, nonmarine sandstone and conglomerate beds of the upper Miocene (6 to 10 m.y.) Mint Canyon Formation. This relationship can best be viewed by taking the off-ramp and then driving slowly up the on-ramp to get back on the freeway.

Inbound: Spotting the sand and gravel operation on the left in the Santa Clara River is easy. The Pleistocene gravel-Mint Canyon Formation angular unconformity is seen in the right-side road cut just before the Soledad Canyon overpass.

● In the next road cut on the left, 0.3 mile north, is another exposure of the Pleistocene iron-stained gravels. Just beyond and about a mile off at 2 o'clock is the mouth of Soledad Canyon. The outcrops there are crystalline igneous rocks. Soledad Canyon carries the runoff of most of the northwest flank of San Gabriel Mountains, which compose the high skyline to southeast. You are traveling a course that has already circled the west end of the range.

Inbound: The iron-stained gravels are in road cuts on the right and Soledad Canyon is to the left as you approach Shadow Canyon Boulevard.

● For the next 3.5 miles, the road cuts and the hillsides provide exposures of Mint Canyon beds, all tilted at a modest angle. Some of the more massive, better cemented sandstone and conglomerate layers make cliffs, bluffs, and ledges on the hillsides.

Inbound: You are still in Mint Canyon beds at "Elevation 2,000 Ft."

These beds and the similar but locally coarser and more reddish layers of the underlying Vasquez Formation, constitute an accumulation exceeding 20,000 feet in thickness, laid down rapidly within the Soledad Basin. At times the ocean lay along the western edge of this basin but never penetrated far into it. The bordering country, particularly to the southeast, was high standing and episodically shed great quantities of very coarse, broken-up rock into the basin. At other times conditions were quieter, and finer sandy and silty beds were laid down along streams and in shallow ponds. A major part of the sedimentary deposition was preceded by extrusion of several thousand feet of lavas, which you will see shortly. The Mint Canyon beds

have yielded remains of fossil horses, camels, and rodents, but no vertebrate fossils have been found in the Vasquez Formation.

Inbound: You are into the Mint Canyon Formation beyond Agua Dulce Canyon. Before that you saw good exposures of the Vasquez at 3 o'clock.

● At Agua Dulce Canyon you can exit and take a short drive left to Vasquez Rocks, a county park among picturesque red *hogback* ridges formed by erosion of resistant sandstone and fanglomerate layers in the Vasquez Formation (Photo J–3).

Inbound: Turn off right and follow signs to Vasquez Rocks County Park.

④ In the first road cut on the left beyond the Agua Dulce intersection, grayish-brown, coarse, bouldery fanglomerate layers rest on pink fine siltstone and sandstone layers. This is the approximate contact of the Mint Canyon on the Vasquez Formation that shortly becomes much coarser. Note the *cavernous weathering* in sandstone ledges on hill slopes beyond this exposure.

Inbound: This contact is on the right as you drop down to the Agua Dulce Creek.

● By looking left at 9 o'clock through a gap 1.6 miles beyond Agua Dulce Canyon, you catch a quick glimpse of Vasquez Rocks (see Photo J–3).

Inbound: This good view of Vasquez Rocks is at 3 o'clock, 1.5 miles beyond Escondido Canyon Road exit.

● In another half mile dead ahead are dark volcanic hills. These are part of a thick section (4,500 feet) of lava flows within the lower Vasquez. Two and one-half miles beyond Agua Dulce Canyon a deep, double-walled road cut gives good exposures of the lavas. Immediately succeeding road cuts and the intervening slopes are also in lava.

Photo J–3.
Vasquez Rocks north of Highway 14 beyond Agua Dulce Canyon, created by differential erosion of reddish sandstone and fanglomerate beds.

Inbound: You come to the road cuts with volcanics 0.5 mile beyond Escondido Canyon Road exit, and they continue for 0.8 mile.

● Just beyond the Escondido Canyon exit the brown ridge to the left at 9 o'clock, with an airway beacon on top, is underlaid by a granite-like rock called syenite, which has been dated as 1,220 m.y. old (Precambrian). To the right at 2 o'clock on the skyline, dipping fanglomerate beds are seen in the Vasquez Formation underlying the lavas.

Inbound: Approaching Escondido Canyon, the ridge with airway beacon is at 3 o'clock.

⑤ One mile beyond the Escondido Canyon exit is a large whitish area on the lower hill slopes just southeast of the freeway, where considerable grading has been done. It is underlaid by fanglomerate beds containing huge fragments of an unusual pale igneous rock called *anorthosite,* which composes bedrock exposures within the San Gabriel Mountains to the southeast. Anorthosite is a rare rock made up mostly of large crystals of calcium-rich feldspar. It is of particular interest because a major component of the uplands of the

moon is anorthosite. Extensive exposures of it are seen along Mill Creek on the Segment I trip, wherein it is more completely described.

Inbound: This locality is about a mile beyond Escondido Summit.

● Just beyond Escondido Summit (3,258), the Truck Parking Area is a good place to stop. The rocks here are Precambrian igneous and metamorphic rocks from which a small production of gold has been obtained. The high, far skyline peak at 12:15 o'clock (snow covered in winter) is Mt. Williamson (8,214). The skyline ridge at 9 o'clock is Sierra Pelona, home of the famed Sierra Pelona Formation, a distinctive metamorphic schist. The open valley ahead is partly filled with alluvium and represents a placid landscape that is about to be dissected by tributaries of the Santa Clara River eating headward into it.

Inbound: Stop in the truck parking area approaching Escondido Summit to see these features.

● One mile beyond Red Rover Mine Road you get a view of other high peaks within the San Gabriels, Pacifico Mountain (7,124) at 1 o'clock, and Mt. Gleason (6,502) at 2 o'clock.

Most of the face of the San Gabriels, as viewed from here, is composed of Lowe granodiorite, an intrusive igneous rock about 220 m.y. old, extensively seen on the Segment I trip.

Inbound: High San Gabriel peaks are seen to the left as you approach Red Rover Mine Road.

● About 0.7 mile beyond Crown Valley Road exit, cuts on both sides of the freeway expose more lavas, part of a down dropped fault block.

Inbound: These volcanics start at "Santiago Road Exit 1 Mile" and extend for 0.3 mile.

● One-half mile beyond Soledad Canyon Road exit is a large road cut on the left exposing a complex of igneous rocks of the Lowe granodiorite group; here badly fractured, shot full of dikes, and deformed.

Inbound: This road cut is on the right near "Soledad Road." In another 1.5 miles is a good view into the wide open valley around Acton.

● Soon you come to the Pearblossom-Angeles Forest exit and shortly thereafter make junction with field trip Segment I, which has come across the mountains from La Cañada. Proceed north on Segment K, or return to Los Angeles via Segment I.

Segment K—Vincent Junction (Palmdale) to Mojave, 40 Miles

● Beyond the junction with Segment I (Figure 3–15), Antelope Valley Freeway (Highway 14) passes through five successive ridges in deep, double-walled road cuts. Cuts 1, 3, and 4 expose tilted layers of deeply altered and weathered lavas, or lavas and volcanic *agglomerates.* Cut 2 exposes fanglomerates and agglomerates, and cut 5 is in old crystalline rocks.

Inbound: The first cut (cut 5) is entered 0.8 mile from the California Aqueduct.

● The Lamont Odett vista point at 1.8 miles from junction of Segments I, J, and K is worth a stop. Here one sees the western Mojave Desert, Palmdale reservoir, California Aqueduct (officially, the Edmund G. Brown Aqueduct), and the San Andreas rift (Photo K–1). On the 11–12 o'clock skyline are Tehachapi Mountains, and the 3 o'clock skyline displays typical buttes and knobs of the western Mojave. Not all points identified on the Landcaster-West Rotary Club bronze plaque are visible from here. At the north end of the vista-point parking is a bronze dedication plaque to Lamont "Monty" Odett. There used to be a bronze plaque here stating that the 1857 'quake occurred in 1957, but it has judiciously been removed.

Palmdale Reservoir lies in a vale created by erosion of the soft, ground-up rocks along the San Andreas rift, here about 1 mile wide. This vale continues northwestward as Leona and Anaverde valleys. The most recent (1857) line of displacement with about 20 feet of right lateral offset, lies along the far side of the reservoir. It crosses the freeway just this side of the ridge through which the freeway passes in a double walled cut. The ridge is a slice within the fault zone.

The old (1913) Los Angeles-Owens Valley Aqueduct crosses the San Andreas fault underground near Elizabeth Lake, 15 miles northwest, but the newer California and Los Angeles-Owens Valley (1970) aqueducts cross on the surface so that repairs can be made in case of a fault displacement.

Inbound: Lamont Odett vista point is not accessible to south-bound travelers.

● Back on the freeway you cross the California Aqueduct. Then 0.4 mile beyond the next exit (Avenue S) enter the double-walled road cut. After crossing the Avenue S underpass

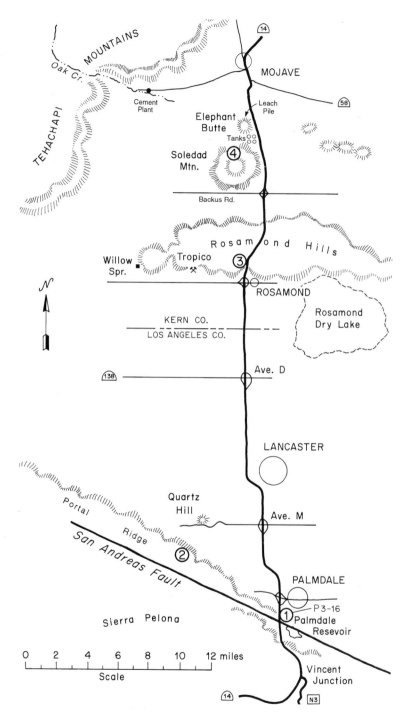

Figure 3–15.
Segment K, Vincent Junction to Mojave.

Photo K–1.
Foreground California Aqueduct, midground Palmdale Reservoir, far side of reservoir is along most recent line of displacement within San Andreas fault zone. Light spot on ridge, left margin, is double-walled Anaverde cut. (Photo by Helen Z. Knudsen.)

note that the freeway changes from concrete to asphalt for 0.4 mile. When first completed, it was all concrete, but in short order bumps developed in the concrete right where crossed by the most recent line of San Andreas activity. Initially, the bumps were attributed to current deformation on the fault, but subsequently it was realized that they were more likely the result of inhomogeneities of the macerated rocks of the fault zone. The entire section through the cut was repaved with asphalt that is easier to repair. You should be able to feel some minor ups and downs even today.

① In cuts on both sides, note the severe folding and crumpling of shale and siltstone beds within the Plio-Miocene (5 to 8 m.y.) Anaverde Formation (Photo K–2). These beds are rich in gypsum ($CaSO_4\cdot2H_2O$) and in early days were mined, probably for agricultural use. To geologists this has become known as the Anaverde cut.

Inbound: The Anaverde cut in San Andreas fault zone is entered one mile south of the overpass at Palmdale Blvd. The contrast of concrete versus asphalt is obvious; small bumps are felt southbound as easily as northbound. Look at the walls of the cut.

● As the freeway clears the low hills, you enter Antelope Valley, a flat, largely alluvial area. On the eastern skyline are low residual bedrock peaks, ridges, and rounded domes greatly reduced by erosion. Westward the country is underlaid primarily by a deep, unconsolidated alluvial fill, mostly 1,000 to 2,000 feet thick but locally exceeding 5,000 feet, as shown by deep wells.

② A little north of Palmdale (just beyond Avenue P) look back at 7–9 o'clock to see the mountain front along the southwest edge of Antelope Valley as defined by strands of the San Andreas fault. The forelying Portal Ridge (see Photo 2–8) at 9 o'clock is a fault slice on this side of the San Andreas. Lighter colored Ritter Ridge at 7 o'clock is of similar origin. The high skyline ridge behind at 7–8 o'clock is Sierra Pelona. It is composed of metamorphic rocks appropriately named the Pelona schist, better known to laymen as Bouquet stone from

Photo K–2.
Crumpled beds of gypsum-rich Anaverde Formation in fault-slice ridge of San Andreas fault zone, exposed in road cut of Highway 14 near Palmdale.

quarries in Bouquet Canyon. This is one of the more widely used southern California building stones.

Inbound: If visibility is good, Portal Ridge has been in view since you left Rosamond. Sierra Pelona rises behind Portal Ridge.

● Starting about 1.1 miles north of Avenue P, at Call Box 624, the freeway enters an entrenched course, which extends 3 miles to Avenue L. Ask yourself why in the world the engineers entrenched the freeway here: to avoid the wind? Prevent noise disturbance? Frustrate snipers? The answer, we think, is that they needed dirt, lots of dirt, to elevate freeway overpasses above city streets in the Lancaster-Palmdale area, and this was a convenient place to get it.

Inbound: At Lancaster note that much of the freeway is continuously elevated between Avenues J and I, so city streets pass under. This elevated position is made possible by dirt from the entrenched freeway section.

● This is a good time to relax. You see little geology close at hand until the highway climbs into hills near Rosamond. If the weather is clear, north of Lancaster you get a sense of the wedge shape of the western Mojave as defined by convergence of San Andreas fault zone behind and Garlock fault zone ahead (see Photo 2–8). The point of the wedge is 30 miles west at about 9:30 o'clock.

Tehachapi Mountains make up the far northwest skyline, and the high peak at about 10:30 o'clock from the 20th Street overpass is Double Mountain (7,988). The Tehachapis as seen from here are composed primarily of granitic rocks enclosing older metamorphics that contain the marble beds that furnish carbonate rock for large cement plants in the Tehachapi-Mojave area. The half-mile long rectangular basin, often water filled, right of the freeway just beyond Avenue H was presumably excavated to get dirt to raise side-street overpasses.

Inbound: Travelers looking south get good views of the San Gabriel Mountains, spectacular when snow capped; Tehachapi Mountains and the Double Mountain are seen on western skyline.

> ▶ Note your odometer reading at Kern County line, Avenue A.

● The Rosamond Hills are dead ahead. Their linear south margin is defined by the Rosamond-Willow Springs fault, and they are composed of granitic rocks with a south-flank mantle of younger deposits, the mid-Tertiary (10 to 20 m.y.) Tropico Group, named from Tropico Mine about four miles west. East of Rosamond the Tropico Group contains much fine-grained lake sediment, but here it consists largely of deformed layers of fine fragmental volcanic debris (tuff), sandstone rich in volcanic particles, volcanic fanglomerates, and lavas with related near-surface intrusive bodies such as plugs, pods, and dikes.

③ Ascending into Rosamond Hills, the smooth white to light-green slopes mark exposures of fine-grained volcanic tuff and tuff-breccia. The darker craggy outcrops are lavas, shallow intrusive bodies, or stony fanglomerate. In road cuts and on adjacent slopes are rocks of various colors—white, cream, green, red, and brown in various shades—formerly quarried for roofing granules. Such a combination usually indicates the presence of considerable fragmental volcanic debris.

Inbound: These volcano-clastic sediments and rocks are exposed in road cuts where the freeway curves and descends to Rosamond.

About 30 miles east-northeast near Boron, the Tropico Group contains the world's richest known deposits of borate minerals (see Mojave Desert province, resources, and field trip Segment G). One of the abundant minerals is *kernite*, named in recognition of this famous locality in Kern County.

Inbound: The backside of the Rosamond Hills looms ahead. The broad featureless area crossed to reach them is underlain by coarse-grained granitic rock that disintegrates readily.

④ About seven miles beyond the county line the highway straightens, and Soledad Mountain (4,183) looms up at 10:30 o'clock (Photo K–3). In times past, Soledad Mountain and Elephant Butte, the smaller mountain northeast of it (which you will see just beyond Mojave's huge water tanks) were the sites of productive gold and silver mines, the famous Golden Queen and Silver Queen. Many abandoned mine workings dot the lower slopes of these mountains, the waste piles of the Golden Queen being prominent on the north slope of Soledad Mountain. Total production from the district amounted to roughly $26 million.

Inbound: Soledad Mountain and its mine workings have been clearly in view since you left Mojave and are seen from many points looking right as you proceed south.

● Soledad Mountain is basically a cluster of intrusive plugs, pods, and dikes of fine-grained, reddish-brown igneous rocks called *rhyolite*, macerated examples of which are exposed in the road cuts on either side 1.4 miles beyond Backus Road overpass. These bodies were intruded into older volcanic rocks, probably in Miocene (5 to 23 m.y.) time, and the ore mineralization is thought to be related to this activity. Fine-grained intrusions have usually been emplaced near the surface, where they cooled quickly. The associated ore deposits are sometimes fabulously rich but normally of no great depth. Mining here extended down only 1,000 feet.

Inbound: The road cuts in rhyolite are 1.6 miles beyond Silver Queen Road overpass.

● The low knobs and ridges east of the highway here (Photo K–4) are, in large part, similar small intrusive plugs etched out by erosion.

Inbound: These are easily seen to left from top of railroad overpass at Mojave and beyond.

Photo K–3.
Soledad Mountain, from Backus Road, composed of near-surface rhyolitic intrusions into mostly Miocene volcanics, site of early mines.

Photo K–4.
Typical Mojave Desert buttes east of Soledad Mountain. Erosion residuals on local rhyolitic intrusions or capped by resistant lavas. (Photo by Helen Z. Knudsen.)

● From *"Silver Queen Road 1 mile"* towers of the extensive windmill farms, far west of Mojave, are coming into view at 10–11 o'clock. The four huge tanks on the left, 0.6 mile farther, house Mojave's water supply. To the right across the railroad tracks before Silver Queen Road is a carbon-black plant, which uses gas from a large pipeline passing through here. Such plants make unpleasant odors, and this part of Kern County is zoned for them. The frequent strong winds help. Elephant Butte is in good view at 11–11:45 o'clock just beyond Silver Queen Road.

Photo K–5.
Leach pile of gold mining operation at Elephant Butte near Mojave.

Inbound: Carbon-black plant on left by railroad opposite Mojave water tanks.

● About 1.5 miles beyond Silver Queen Road overpass, at 9 o'clock in front of the close hills, are two huge leach piles (Photo K–5) of a gold mining operation (Bexus Rock Products). Rock (ore) quarried from the nearby hills is crushed, put into the pile, and then sprinkled with a cyanide solution that leaches the traces of gold in the rocks. The solution is collected on an impermeable membrane underneath the pile, and the gold is recovered by precipitation. This procedure has revolutionized and rejuvenated mining in many old, abandoned gold districts, like Mojave, in the west. The leach piles look like the beginnings of huge pyramids.

Inbound: The leach piles are at 1 o'clock beyond Purdy Road.

● Beyond Elephant Butte, look west at 9:15 o'clock in passing Purdy Road to see, seven miles off, the dust from a large cement plant and the light-colored areas on the hills behind, if the afternoon sun is not in your eyes. The light areas are the quarries from which marble rock

for cement is obtained. Windmills are in clear view to west between the leach piles and Purdy Road.

Inbound: Look right at Purdy Road for the cement operation.

● As you come into Mojave, the lower, more subdued skyline at 10–11 o'clock indicates the location of upland Tehachapi Valley which separates the Tehachapi and southern Sierra Nevada mountains.

Inbound: From overpass across railroad south of Mojave, you get good views of Soledad Mountain and old mine workings at 12:15 o'clock, of leach piles at foot of Elephant Butte at 12 o'clock, cement plant operations far west at 3 o'clock, and wind farms on skyline at 3:30 o'clock.

Segment L—Mojave to Junction with U. S. 395, 47 Miles

You leave Mojave headed north-northeast on Highway 14 toward Bishop (Figure 3–16).

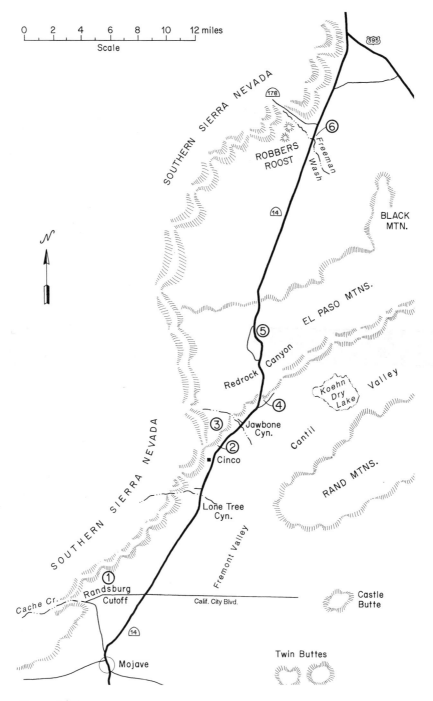

Figure 3–16.
Segment L, Mojave to Highway 395.

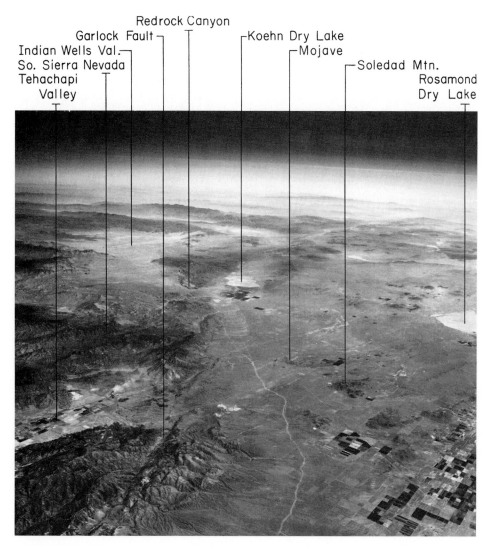

Redrock Canyon
Garlock Fault
Indian Wells Val.
So. Sierra Nevada
Tehachapi
Valley
Koehn Dry Lake
Mojave
Soledad Mtn.
Rosamond
Dry Lake

Photo L–1.
High-altitude oblique view looking northeast along trace of Garlock fault, mountains on one side, low relief desert on the other. (U.S. Air Force photo taken for U.S. Geological Survey 064L-093.)

▶ Record odometer reading at the Y inter-section of Highways 14 and 58 on the north edge of town.

Our route north-northeast to Redrock Canyon parallels the south end of the Sierra Nevada, which is cut off abruptly by the Garlock fault zone (Photo L–1). To the right going out of town is Mojave Airport, home of the nonstop globe-girdling plane, "Enterprise." Surplus commercial jumbo jets for sale are now stored here. This is a used-plane lot for aircraft.

Inbound: The airport is easily seen as you come into Mojave.

Cache Creek
(Garlock Fault)

Mojave

Highway 14

Forelying Fault Block

Hwy. Intersection
So. Sierra Nevada

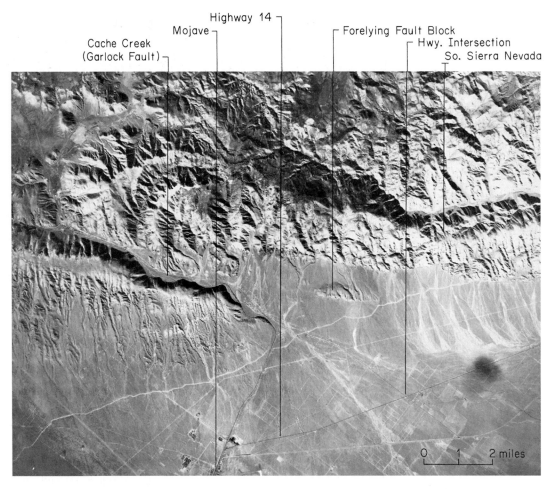

Photo L–2.
High-altitude vertical air photo just north of Mojave, north toward upper right corner, scale at lower right. (U.S. Air Force photo taken for U.S. Geological Survey, 744V-031.)

① At 9:00 o'clock beyond the first curve, about 2.7 miles from town, is a flat-topped ridge with a steep south face, fronting higher background mountains. It is a forelying fault block (Photo L–2) lifted along one of many parallel faults within the Garlock zone. To the left at about 8 o'clock the railroad and highway to Tehachapi turn west-southwest just within the mountains to follow up Cache Creek, which has carved its canyon along the Garlock (Photo L–2).

Inbound: This fault block is seen well at 1 o'clock, 2 miles south of Phillips Road turnoff.

California City Boulevard turns right five miles out, and a mile or two beyond on the 3 o'clock skyline is Castle Butte, more of a ridge than a butte. The two separate smaller conical knobs at 3:15 o'clock are Twin Buttes. These prominences are mostly plugs of fine-grained intrusive rock, etched out by erosion, or knobs capped by a protective lava flow. The broad, smooth surfaces between are either alluvium or exposures of granitic rock worn smooth by weathering and erosion. In clear weather with low-angle lighting this is a striking landscape.

Photo L–3.
Lava flows cap softer sedimentary beds of upper part of Ricardo Group, Red Rock Canyon.

Inbound: Views of these buttes are good as you approach California City Blvd.

● Along the straight highway beyond, you drive directly toward El Paso Mountains on the horizon. Skyline hills at 1:30 o'clock are part of the Rand Mountains, site of the historic mining camps of Randsburg and Johannesburg farther east (see field trip Segment G).

Inbound: The Rand Mountains have made the eastern skyline since Red Rock Canyon.

The relatively smooth surface being traversed is an apron made up of coalescing alluvial fans spreading outward from the southern Sierra Nevada. The fans, composed of fine-grained disintegrated granitic debris, are smooth and gently sloping. You see only occasional cobbles and pebbles. The southern Sierras are made up mostly of granitic rocks, with thin slices of metamorphics along the range front imparting variegated colors. Tertiary sedimentary and fragmental volcanic materials crop out farther back in range.

Inbound: You traverse this area between Lone Tree Canyon and the California City Boulevard. Soledad Mountain guides you toward Mojave, and good views of Castle and Twin buttes have been yours since the Randsburg Road.

● The next seven miles are relatively uneventful, so information on the Ricardo Group (extensively exposed ahead in Redrock Canyon) is inserted.

The Ricardo is a terrestrial (land-laid) deposit nearly 7,000 feet thick that accumulated in a local land-locked basin during the Miocene (7 to 19 m.y.) time. The formation consists primarily of layers of sand, silt, clay, and gravel, all containing considerable fragmental volcanic debris. Variegated colors in parts of the Ricardo reflect this volcanic content. Also interbedded are two groups of dark lava flows, which cap prominent ridges that are etched out by erosion (Photo L–3). Notable near the center of the Redrock amphitheater is a thick, massive, delicate

Photo L–4.
Massive, jointed, pink tuff bed lying above well-stratified upper Ricardo Group deposits,
Redrock Canyon.

pink layer of volcanic tuff-breccia (Photo L–4) formed by eruption of hot volcanic ash and other fragmental volcanic debris.

The Ricardo deposits are soft; hence they are easily and rapidly eroded. Yet some of the layers are coherent enough to stand in vertical faces. This property of easy erosion and large differences in the resistance of layers, helps create the castellated forms (Photo L–5), spires, chutes, niches, alcoves, and other badland features that, along with variations in color, give Redrock Canyon its striking character. A modest northwestward inclination of the eroded layers adds variety to topographic forms. Many people the world over have seen Redrock Canyon in the background of western thrillers and television advertisements, because this colorful and unusual terrain has long been fancied by movie and television companies.

The Ricardo beds are fossiliferous, and petrified wood is collectible in places. Harder to find are the bones and teeth of vertebrate animals that inhabited the Ricardo Basin millions of years ago. They composed an interesting assemblage including horses, camels, mastodons, rhinoceroses, wild dogs, pronghorn antelope, deer, saber-tooth tigers, smaller cats, weasels, rabbits, and a goose. These animals lived in an open forest-grassland environment with perhaps 15 inches annual rainfall—roughly three times the present quota. Among the trees were live oaks, piñon pine, locust, cypress, acacia, and palms. You should look forward to seeing the Ricardo.

● Eleven miles out, Phillips Road turns off right and 0.5 mile beyond is "*Passing Lane 2 miles.*" In another 0.7 mile the highway curves left and a little beyond, left of the highway, within the alluvial apron, is a somewhat bare,

Photo L–5.
Gently tilted beds of the Ricardo Group in the castellated Red Cliffs of Red Rock Canyon amphitheater. Pink tuff at top of highest point.

low ridge bearing power line towers. You are looking at the backside of a little fault block uplifted along a strand of the Garlock fault cutting away from the mountain front. The north side of this block is a fault scarp facing toward the mountains. This feature continues in view for nearly 2 miles and approaches the highway until it is terminated by Lone Tree wash beyond Call Box 292 and near the start of a four-lane highway segment.

Inbound: This fault-feature relationship is better seen at 3 o'clock inbound as you cross Lone Tree wash about at the end of the four-lane passing zone, 2.3 to 2.5 miles beyond Cinco.

● Near Cinco, nearly 16 miles from Mojave, the highway curves right and from there to the mouth of Redrock Canyon you travel close along the mountain front. Here the mountains are bounded by El Paso fault, one of the principal strands of the Garlock zone.

Inbound: Cinco was burned out in early 1990s, but scars remain.

② About 0.7 mile beyond burned-out Cinco, the highway curves east parallel to the mountain front, and a discerning observer can begin to see El Paso fault where it crosses gullies and spurs at the mountain base. The forelying spurs are lighter in color, softer looking, and locally show layering. The steep, darker mountain front behind consists of highly fractured, hard, fine-grained, granitic rock.

Inbound: These spurs are on the right 0.3 to 0.7 mile beyond the Sierra fault.

③ In another mile, highly colored rocks compose the mountain front. These are part of the Ricardo Group, and their sudden appearance is caused by a north-trending fault that drops them against older somber rocks to the west. This fault is considered by some geologists to be the southern end of the Sierra Nevada frontal fault (Photo L–6), here terminated at El Paso fault. Eastward the forehills are lower and less rugged, because they are underlain by softer Ricardo rocks.

Inbound: You pass the truncated end of the Sierra fault 0.7 mile beyond the Cantil post office.

● A look at 8 to 9 o'clock, about 0.3 mile beyond the Cantil post office, shows the crossing of Jawbone Canyon by the newer, zig-zag 1970 (white), and much farther back older, 1913 (black), Los Angeles Aqueduct lines.

Photo L–6.
View west-northwest of south end of Sierra Nevada frontal fault, 1.7 miles northeast of Cinco. Somber rock to left largely granitic, variegated rocks to right are younger Ricardo Group beds. Fault plane inclined right. (Photo by Helen Z. Knudsen.)

Inbound: Within 0.4 mile beyond Randsburg Road, Soledad Mountain behind Mojave looms at 11:45 o'clock, and 0.1 mile beyond Rogers Road, the old aqueduct is visible up the Jawbone Canyon.

● In another 1.3 miles the Red Rock-Randsburg Road branches right. For the next 2 miles, approaching Redrock Canyon, you pass through slightly dissected, pale, fine-grained, gently tilted Ricardo beds locally capped by stony gravels. About 1.5 miles beyond the Randsburg turnoff at 10 o'clock the mountain front displays nice badland topography in highly colored Ricardo beds. The band of black rock capping a ridge to the west is one of the lava flow units (see Photo L–3) within the Ricardo.

Inbound: At and beyond "Campers, Trailers, Windy Area," toward 3 o'clock is the badland exposure of Ricardo beds on the face of El Paso Mountains. The wide basin ahead and south is Fremont Valley.

④ A little before you enter the narrow mouth of Redrock Canyon, as the highway curves left, a quick look right at 2 o'clock shows a sharp contact between north-tilted Ricardo beds and crystalline rocks. This is El Paso fault.

Inbound: Southbound this view is more difficult, at between 8 and 7:30 o'clock.

● Within 0.3 mile you enter the narrow, steep-walled, hard, granitic-rock gorge of lower Redrock Canyon. It contrasts markedly with the broad amphitheater carved into softer Ricardo beds, which will open suddenly before you in 0.8 mile as you cross the large concrete bridge.

Inbound: Southbound, the gorge is entered beyond the bridge.

● Right of the highway within the amphitheater is the spectacular castellated and fluted face of Red Cliff (see Photo L–5), accessible

Photo L–7.
Smooth surface truncating tilted strata of upper Ricardo Group is a pediment. This view is best seen inbound descending into Red Rock Canyon. (Photo by Helen Z. Knudsen.)

from the large parking area entered 0.4 mile from the bridge. As the highway curves around the west end of Red Cliff, you will see a massive, pink, volcanic tuff-breccia bed on the left, and you pass through it in 0.1 mile (see Photo L–4).

Inbound: The entrance to the parking area is on the left 0.3 mile beyond the south end of Abbott Road.

● Beyond, to the left is another castellated cliff capped by dark lavas (see Photo L–3). You cross these lavas in about 1.2 miles, but before that, note the wind-blown sand heaped on hill slopes left of the highway.

Inbound: In the next 0.9 mile the highway crosses both sets of lavas, then parallels castellated cliffs to the right and pink tuff to the left before coming to the south end of Abbott Drive. In 0.1 mile more it crosses through the pink tuff, and in 0.2 mile the Red Cliff parking area lies left.

If you wish a more leisurely trip through Redrock, turn off left just beyond the pink tuff bed onto Abbott Drive, and follow this old

two-lane paved road to a junction with Highway 14 about 2.7 miles ahead. Park ranger station, visitor's center, and campground are along this road.

Inbound: The turnoff onto the north end of Abbott Road is on the right 0.3 mile south of "Divided Highway 1 Mile."

⑤ As you climb out of Redrock Canyon on Highway 14, everyone except the driver can look back to 8 o'clock at *"Elevation 3000 Ft."* and see the tilted Ricardo beds sharply truncated by a smooth erosion surface sloping gently eastward (Photo L–7). This surface is what geologists call a *pediment.* It was cut by streams flowing southeastward out of higher mountains to the west. Subsequently, the entire El Paso Range was uplifted and dissected to create the present topography, the pediment being partly destroyed in the process. Southbound travelers get a good view of these relationships just before dropping into Redrock Canyon.

Inbound: Southbound pediment views are best at 12:10-12:20 o'clock near the start of the divided highway.

● About 5 miles beyond the Redrock amphitheater you start another section of four-lane highway, only 1.1 miles long. As you top out at its end, a long straight stretch of highway extends ahead. The next significant curve is nearly 3 miles beyond the horizon, where the highway disappears. Why not have everyone guess how far it is to that curve?

▶ | Note your odometer reading leaving the four-lane section.

Inbound: A good game only outbound.

● As you move onto the straightaway, Black Mountain, so named for an obvious reason, is at 2 o'clock. The black rocks are basaltic lavas, their normally dark color enhanced by a coat of rock (desert) varnish.

Inbound: Black Mountain is well seen at 9:15 o'clock near start of previous passing lane and beyond.

● The long straightaway traverses an alluvial apron composed largely of disintegrated granite derived from the Sierras. Note the sparsity of boulders. In about 3 miles you begin crossing a succession of widely separated, shallow, flat-floored gullies cut into this apron. The best one is in the middle of the next passing lane, 6.5 miles out on the straightaway. The dissection of any alluvial surface is an indication of some change in regiment of the streams flowing across it. In this instance, gentle uplift or perhaps a change in climate has caused the shift from deposition to dissection.

The southern Sierras on the left, lower and less rugged than farther north, are composed of granitic igneous rocks with some metamorphic pendants. For the most part, these rocks weather readily, so the range front is subdued. Robbers Roost (Photo L–8), a small local body of resistant rock, is in good view at 10:30 o'clock and beyond.

Inbound: Best gully is midway through passing lane starting five miles from Highway 78 Junction.

⑥ In another 4 miles you cross a deeper and larger gully identified as Freeman Wash by the sign at the bridge. Note the bouldery layers in its banks 100 feet to the right and large boulders on its floor. These bouldery deposits came from the mountains in a mass of muddy debris, something like freshly mixed concrete, called a *debris flow*. As you rise out of the wash look left at 8:45 o'clock to the mountain front where the forelying isolated ragged knobs of Robbers Roost (see Photo L–8) claim attention. These are not something pushed up out of the ground, but rather residual knobs left as the mountains, of which they were once a part, retreated westward under the attack of weathering and erosion.

Within a mile is *"Highway 178, Isabella Left 1 Mile,"* and 50 feet beyond is Father Crowley's memorial monument.

Inbound: Crowley monument is 0.5 mile beyond Highway 178 Junction, and Freeman Gulch is about 1 mile farther. Robbers Roost (see Photo L–8) is in good view all along from 2 o'clock on.

● A gentle curve in the highway 0.6 mile before Isabella junction hardly qualifies as breaking our straightaway. At and beyond the Isabella turnoff you get good views, in decent weather, of Indian Wells Valley ahead and to the right (see Figure 3–16). The mountains on the far side to the northeast are the Argus Range, the visible part being largely granitic. Their lower slopes are made pale by a mantle of wind-blown sand and silt swept up from the valley. Between 10,000 and 20,000 years ago the eastern and central parts of Indian Wells Valley held a shallow lake that eventually

Photo L–8.
Robbers Roost, erosion residuals of Sierra Nevada granitic rock, seen to the west from Highway 14 in vicinity of Freeman Wash.

merged with a larger and deeper body in Searles Lake basin farther east (see Basin Ranges province, special features, Figure 2–14).

Inbound: Good views of Indian Wells Valley have been available to the left since you left the 395 intersection.

> ▶ Note your odometer reading at the first distinct curve just beyond the Inyokern turnoff and see how closely you guessed the length of the straightaway.

Inbound: The distance guess is no game in this direction.

● In a mile you pass Homestead and Indian Wells, climb another little hill, and as the highway straightens out see far ahead at 10 o'clock volcanic cones and dark lava flows on the south end of Coso Range behind Little Lake (Figure 3–17). These continue in view as you drive north.

Inbound: You have already seen the volcanics of the Cosos.

● Shortly you intersect U.S. 395 coming from the right and join those who elected to start from the San Bernardino-Cajon Pass route via Segment A. At 9:15 o'clock in the core of the Sierra is high, bare, rugged granitic Owens Peak (8,453).

Segment M—Highway 395 Junction to Olancha, 41 Miles

> ▶ Before you drive north from the highways 395/14 junction (see Figure 3–17) record your odometer reading, then keep an eye on the Coso Range ahead at about 12–1 o'clock.

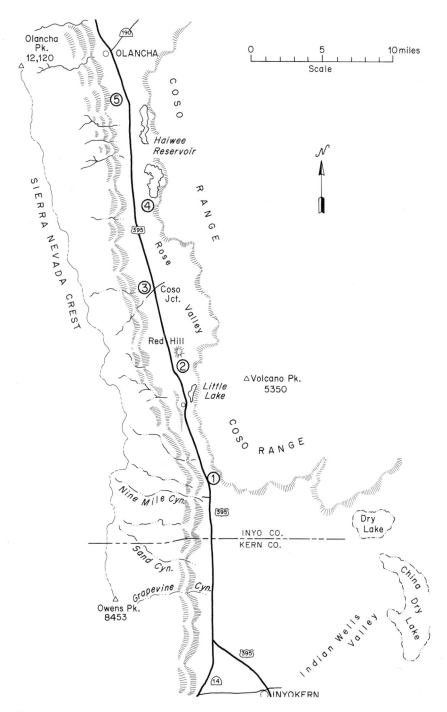

Figure 3–17.
Segment M, Highway 395 junction to Olancha.

● The volcanic cones and flows of dark lava thereon become more apparent on closer approach. This stretch of highway was being converted to four lanes in 1993.

Inbound: These cones and flows are partly in view behind to the left after you leave Little Lake.

● The Sierra Nevada front to the left, now closer at hand, becomes higher and steeper northward to Lone Pine. Here it is relatively subdued, partly because rocks at the range front are older granitics, with some metamorphic pendants, and weather readily so the slopes are mostly debris mantled rather than bare rock. About 2.5 miles from the 395/14 junction, Owens Peak (8,453) is at 9:15 o'clock up Grapevine Canyon. At the core of the range there is hard fresh granitic rock that makes more rugged features, like the range front behind Olancha.

Inbound: You get a good view of Owens Peak at 3 o'clock 1.5 to 2 miles beyond Sand Canyon, with its gravel pit, opposite Brown Road.

▶ Record your odometer reading at Inyo County line.

● In 2.2 miles at 9:20 o'clock, up Noname Canyon, you see both aqueduct lines.

Inbound: Noname Canyon is 1.3 miles south of Kennedy Meadows turnoff.

Beyond the Kennedy Meadows turnoff are exposures of lavas a little right of the highway that will be with you all the way to Little Lake. Just beyond "*Little Lake 7, Independence 71, Bishop 112,*" is a large bare spot on the right, the site of a former lumber mill. On the northeast skyline at 1–2 o'clock are volcanic cinder cones and lava flows on the slopes of the Coso Mountains.

Inbound: Lavas left of the highway disappear a little before the Kennedy Meadows turnoff, Coso volcanics have already been seen, and the sawmill site is easily spotted.

① At 4.8 miles from the county line and 150 yards left are some low bluffs and road cuts along southbound lanes of the separated highway. You can see that some gravel layers in the cuts are exceptionally bouldery. These coarse deposits have been carried from the mountains as sheets of muddy, bouldery, *debris flows.* Such flows have great carrying power, and huge boulders on alluvial fans farther north, as near Olancha, have probably been transported by this means. Debris flows have occurred repeatedly along the Sierra front in the past and still happen following cloudburst rains in the mountains (Photo M–1).

Inbound: These debris-flow deposits are on the right 4.8 miles south of Little Lake after passing under the large power lines and crossing a concrete bridge and are seen closer at hand than northbound.

● At 5.9 miles is a macadam truck parking shoulder, a good place to stop. Ahead is the narrow Little Lake gap through which the highway, railroad, and aqueduct pass (Photo M–2). The gap and the lava cliffs to the right have been carved by a large stream that once flowed through carrying meltwater from ice-age glaciers of the Sierra. We shall call it the glacial Owens River, which flowed south and then east to feed a chain of large lakes in basins as far east as Death Valley (see Basin Range Province, special features, Figure 2–14).

Beyond the truck turnout, the steep face parallel to the highway on the right for several miles has a lava cliff 30 to 35 feet high at the top. The steep slope below is covered by blocks of lava from the cliff that mantle underlying alluvial gravels. At 2:30 o'clock on the near

Photo M–1.
Highway 395 inundated by a debris flow south of
Little Lake in 1956. (Photo by Pierre St. Amand.)

skyline, where the four lanes of our highway come together, is a nice red cinder cone sitting atop the lava cliff.

Lava is exposed in the floor of the abandoned river channel, but it is a younger flow that ran down the channel after it was cut through the lava cliff. Flat topped, cliff forming remnants of a still younger lava flow are to be seen at 10–11 o'clock along the channel margin to the left at the first turnoff into Little Lake.

▶ | Record your odometer reading opposite the abandoned Little Lake Hotel.

Inbound: Views of the lava cliff with its talus and of younger lavas on the floor of glacial Owens River channel are as good, possibly better, inbound. Watch for the youngest lava-flow

remnants about one-quarter mile west at the south exit from Little Lake. Record your odometer reading opposite the Little Lake Hotel.

● Just beyond is the lake itself, a shallow body, nourished by small seepage springs and occupying part of the old river course.

Inbound: No direction needed.

● Passing the lake, Red Hill, a cone of reddish volcanic *cinders,* looms ahead at 12 o'clock (Photo M–3). Lava cliffs at 2 o'clock, north from the lake, display columnar structures about where the highway first curves left (Photo M–4). They are created by fractures that formed as the lava cooled and are known as *columnar jointing.* They are superbly developed at Devil's Postpile on the Middle Fork of San Joaquin River west of Mammoth. Why the Devil gets credit for creating columnar joints remains a mystery, but he does worldwide.

Olancha Pk.
Fault Scarp
Haiwee Gap
Red Hill
Lavas
Dry Falls
Columnar Jointing
Little Lake
Owens Lake
Inyo Mtns.

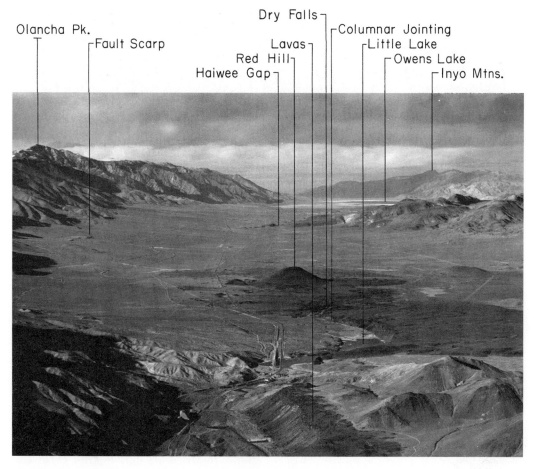

Photo M–2.
Air view north from over Little Lake. (Photo by Roland von Huene.)

Photo M–3.
The cinder cone, Red Hill, just north of Little Lake. (Photo by Helen Z. Knudsen.)

Photo M–4.
Columnar jointing in lavas of east wall of glacial Owens River gorge just north of Little Lake.
(Photo by Helen Z. Knudsen.)

● About 1.5 miles north of Little Lake Hotel at the left curve look at 1:30 o'clock for a quick view of an abandoned gorge cut into the lavas by glacial Owens River. It has a spectacular dry waterfall at its head, with deep pot holes and stream-polished rock surfaces, and is well worth a visit (Photos M–5 and M–6). Good columnar jointing is seen at 3 o'clock from here.

Inbound: Inbound views of lava cliffs, columnar jointing, and the gorge are, if anything, better than northbound at 9–11 o'clock rounding the curve to the left 1.5 miles beyond Cinder Road.

● The scars of mining operations for cinders are obvious on the south flank of Red Hill, after the highway straightens out.

② To visit the dry falls, turn off east on Cinder Road 3 miles north of Little Lake. This junction is marked by an intersection sign, and the turnoff is paved. Follow this road east for 0.5 mile, take the first well-traveled right branch south, follow it 0.3 mile to "*Not a Through Road*" and turn left on the side road. Follow it 0.5 mile to a wide parking area in the lavas.

Orange dabs on rocks at the east edge of the parking area mark a trail east across an uneven lava surface for a few hundred yards, a 10-minute walk. Note that the rocks are full of little holes, 0.25 to 0.5 inch across. These were made by gas bubbles in the molten lava.

The falls are 150 feet right of your first encounter with the old stream course. They occurred in two drops totaling more than 100 feet, the second being about 200 feet farther down stream. Note the rock surfaces scoured, smoothed, and fluted by the fast-flowing,

Photo M–5.
Looking upstream into glacial Owens River gorge containing the dry falls.

Photo M–6.
Downstream view of gorge below upper dry falls. Foreground rock surface scoured, polished, and potholed by the rushing glacial waters. Picnic party near lip of lower falls.

presumably silty waters (see Photo M–6). Pot holes were cut into the rock by fixed vortices in the stream. Water flowed past this locality several times, but the features you see were formed primarily during the last major discharge, about 10,000 to 15,000 years ago.

This was an attractive place then. Indians found it so and spent much time in the area, as shown by nearby living sites and artifacts discovered by archaeological explorations. Small chips of shiny obsidian (volcanic glass) are still scattered along the banks of the river channel here. The obsidian came from outcrops in the Coso Range only a few miles northeast.

Inbound: Cinder Road is just south of Red Hill. It's easy to spot.

The lava flows making the long cliff coming into Little Lake are thought to be about 400,000 years old. The older flow in the glacial Owens River channel is possibly more than 100,000 years old, and the younger channel flow is perhaps only 20,000 years old. The columnar jointed lava and the lava at the Dry Fall belong to the 100,000-year group.

● Returning to the highway, signs of the cinder mining operation on the south flank of Red Hill are close by. Volcanic cinders are not the remains of burned rock but rather small, pebble-sized fragments of highly porous lava, usually red or black, explosively thrown out of a volcanic vent. Cones are formed as the cinders pile

Photo M–7.
North side view of Red Hill showing breached crater. (Photo by Helen Z. Knudsen.)

up around the vent. Cinders are extensively used as a lightweight aggregate in concrete and cement blocks.

Inbound: You can see the breached crater of Red Hill (Photo M–7) but not much of the cinder mining operation or the big road cut from Highway 395. Approaching Red Hill note the knobs of lava projecting through their mantle of windblown granitic debris.

● Back on the highway headed north, you traverse knobby lavas partly buried by windblown disintegrated granitic detritus for 2 miles. In 0.6 mile from Cinder Road a big road cut on the east (right) exposes a good cross section of the cinders and lava blocks thrown out of the Red Hill vent. Beyond Red Hill, look to 4–5 o'clock (right) to see volcanic cones and a lava flow descending a canyon in the Coso Range (Photo M–8). A still better place for viewing the cones, flows, and some of the many lava domes of the Coso volcanic field is from the east edge of the rest area ahead, at Coso Junction (population 9). The Coso volcanic area also boasts hot springs, fumaroles,

and mercury deposits. It is the site of geothermal electrical power plants (steam columns). You are seeing only the periphery of a large volcanic area with volcanic features as old as 25 m.y. and as young as 40,000 years. The area is not open to the public, because it is part of the Navy's China Lake bombing range.

Inbound: Southbound views of the Coso volcanics are good from north of Coso Junction to Red Hill at 10–11 o'clock. The cinder cone (Volcano Peak) at 10:30 o'clock on the high skyline with its lava flow running down a canyon is spectacular (Photo M–8). It is about 40,000 years old.

● North of Red Hill the highway enters Rose Valley where the grove of cottonwoods ahead marks the location of Coso Junction (rest stop, gas station, and cafe). Some ponding of glacial Owens River probably occurred in this valley, but this shallow pond did not leave strong shoreline markings or thick deposits. The white area at 9 o'clock a mile or so west of Coso Junction marks the site of a former mill processing pumice (a frothy volcanic glass) from

Photo M–8.
Volcano Peak and lava flow from its base as seen east of Highway 395 north of Red Hill.
(Photo by Helen Z. Knudsen.)

the Coso Range. White areas on the face of the Cosos at about 2 o'clock behind Coso Junction are clay pits.

③ Farther away, at 9 o'clock from Coso Junction near the base of the Sierra, the alluvial apron is broken by a linear embankment, Portuguese Bench, green in summer with grass and scattered trees. This is a fault scarp remnant along a fracture within the Sierra Nevada frontal fault zone. Geologically, this scarp is not very ancient, but it is old enough to have been erased by erosion and deposition opposite the mouths of major canyons.

Inbound: Portuguese Bench is seen at 1:30–2 o'clock at "Rest Area 1 Mile."

④ Beyond Coso Junction near *"Olancha 17, Lone Pine 40, Bishop 98,"* at about 12:15 o'clock, the mouth of a stream-cut gorge is visible at the base of the hills. It transects a bedrock spur projecting west from the Cosos and is the site of the Haiwee dam and power-

house. It was cut by the overflowing glacial Owens River. Keep an eye on this locality as you travel north. The clay pits earlier noted are clearly visible at 12:45–1 o'clock.

Inbound: As you descend into Rose Valley beyond power lines crossing, look back at 7–8 o'clock to see the gorge.

● Shortly you ascend a gentle grade out of Rose Valley, and straighten out at a higher level. At 2:30 o'clock along the base of the Coso Range the scar of a large rock slide is just upstream from the stream-cut narrows (Photo M–9). The slide involved mostly hard brown rock. Its head is marked by a high cliff, and the body is an irregular bulb of huge blocks. It was probably caused through over-steepening of the hill face by the undercutting of glacial Owens River.

Inbound: The breakaway scarp and slide body are seen at 9:30 o'clock from "Rest Area 6 Miles."

Photo M–9.
Rock slide at narrows below Haiwee dam. The cliff is breakaway scarp of slide; the irregular mass below is body of slide.

● At *"Lone Pine 32, Bishop 91, Reno 294,"* water is usually visible in lower Haiwee Reservoir (Photo M–10), part of the Los Angeles Aqueduct system. The reservoir occupies a former meadow bordering the course of glacial Owens River. White rocks in the hills behind are part of the Coso Mountains Formation, a late Pliocene (2.5 to 6 m.y.) land-laid sedimentary deposit, containing fossil remains of large horses, slender camels, hyena-like dogs, short-jawed mastodons, meadow mice, and rabbits. The region must have been a grassland at that time.

Inbound: White Coso Mountain beds make up lower slopes on far side of Haiwee Reservoir.

● Within 2.5 to 3 miles beyond *"Elevation 4000 Ft."* note that fan surfaces along the highway are bouldery, indicating a more durable kind of rock in the Sierra than northeast of Mo-

jave or west of Indian Wells Valley. Fault-scarp remnants adorned with trees and houses are seen to the left, close to the road, 3.3 miles beyond *"Elevation 4000 Ft."*

Inbound: You see this scarp at 4 o'clock, 0.3 mile beyond Sage Flat Road.

● Ahead less than a mile beyond Sage Flat Road the usually dry flats of Owens Lake come nicely into view. Before the 1913 diversion of Owens River into the Los Angeles Aqueduct, this was a large, blue, salt-water lake of considerable beauty owing to the desert-country setting at the base of towering mountains. In wet years water still covers part of the lake floor and may linger a year or two before evaporating. The pre-1913 lake was about 30 feet deep and covered 100 square miles. Glacial Owens Lake that fed the overflowing glacial Owens River was 220 feet deep and about twice as large.

Photo M–10.
High-altitude oblique view south-southwest over Owens Lake bed. (U.S. Air Force photo taken for U.S. Geological Survey, 374R-184.)

Photo M–11.
Sierra Nevada front from Olancha. Massive granitic rocks in range core.

Inbound: You have seen much of Owens Lake already.

In less than a mile, after the highway starts a descent toward Owens Lake the open concrete trough of the Los Angeles Aqueduct crosses under the highway. Here, a few hundred yards left, is an old, degraded, bouldery fault scarp breaking the fan surface. An abandoned railroad bed traverses its face.

Inbound: The north end of degraded bouldery scarp is three miles out of Olancha.

● Coming into Olancha, small dunes visible on the valley floor at about 1:45 o'clock from *"Death Valley, Right Turn 1 Mile,"* have been built by south-blowing winds, largely since

Owens Lake dried up. Note how high and abrupt the Sierra front has become (Photo M–11). Olancha Peak rises to 12,120 feet, but Olancha is at 3,649 feet.

Inbound: As you leave Olancha, the dunes are east on the valley floor at about 8 o'clock within the first mile.

Segment N—Olancha to Independence, 38 miles

● Within the first mile north of Olancha (Figure 3–18) the bouldery surface of the Cartago Creek fan is seen ahead and to the left. The

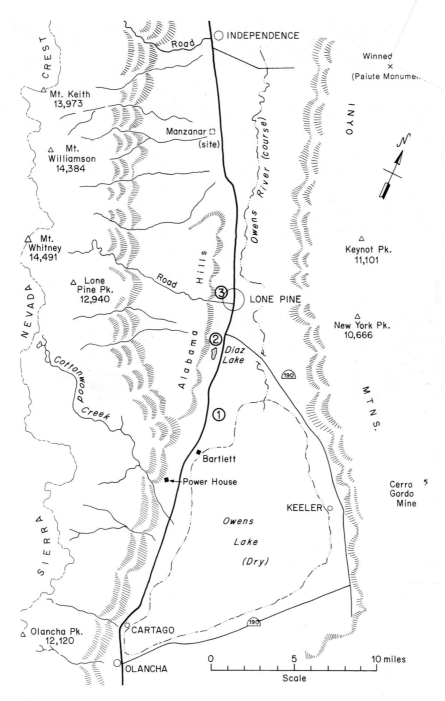

Figure 3–18.
Segment N, Olancha to Independence.

Photo N–1.
Cartago Creek fan as viewed leaving Olancha (northbound), with automobile-size boulders reflecting massive granite in range core (Photo M–11.) (Photo by Helen Z. Knudsen.)

large boulders (Photo N–1), some the size of a small house, result from massive sparsely jointed rocks in the mountains and testify to the transporting power of debris flows. The unusual abruptness and ruggedness of the Sierra Nevada front here (see Photo M–11) reflects the influence of the massive granitic bedrock.

Note the absence of roadside billboards here and on up Owens Valley, compared to the south approach to Olancha; thank the city of Los Angeles, which prohibits billboards on its property.

Inbound: You see these features as you approach and move beyond Cartago.

● Owens Lake was saline, and in the past chemical plants recovered salts from brines and evaporated residues on the lake flats after it became

dry in 1926. The principal compound produced was sodium carbonate or soda ash, which is extensively used in glass making and ceramics. Cartago, just ahead, was once a center of such operations, and white waste piles there mark the sites of former chemical plants. Remnants and waste piles of Lake Minerals Corporation's abandoned chemical plant are seen to the right two miles beyond Cartago where a passing lane begins. Lake Minerals still recovers Trona, a complex sodium carbonate salt from Owens Lake and has plans to expand its operation.

▶ | Note your odometer at Cartago. |

Inbound: A good white waste pile is seen at 11 o'clock coming into Cartago.

● About 2.5 miles north of Cartago, look right to the broad, low, skyline saddle that separates Coso Mountains from the Inyo Range to the north. At 2:30 o'clock is Malpais Mesa, an area of nearly flat dark lava flows dissected by canyons and dropped westward in a series of step faults. The lavas are a veneer on older rocks.

Note that the riparian vegetation along stream courses on fans here ends at the aqueduct. This happens because the aqueduct swallows the streams, and the trees below have died.

Inbound: These features have been in good view across Owens Lake from two miles south of Bartlett.

● The front of the Sierra Nevada begins to change character opposite Cartago, and at Cottonwood Creek, 6.3 miles north, the range front is less abrupt, less rugged, and is partly mantled by loose, weathered debris. This mantle is cut by fresh gullies on the mountain face back of Cottonwood Creek power house, the clump of trees and cluster of buildings 1.8 miles north of Cottonwood Creek. The gullies formed when a flume high on the hillside broke. This range-front difference is caused by a change in bedrock, from massive, clean, Sierra granite at Olancha, to an older mixture of granitic and metamorphic rocks that weathers more easily.

Inbound: The mountain-face gullies are seen at Cottonwood Power House Road, 2.2 miles south of Bartlett.

● Cottonwood Creek (dry) is crossed on a concrete bridge and 0.4 mile farther the road to Cottonwood charcoal kilns turns off east. The charcoal was used at Swansea across Owens Lake to smelter ore from the Cerro Gordo silver mine.

Inbound: Coming south in clear weather, Telescope Peak in the Panamints suddenly appears on the far distant skyline from behind the nearer mountains. It is most easily recognized in winter when the peak is capped with snow.

● After you pass Cottonwood Creek Power House Road, you can look to the eastern skyline of the Inyos at about 2:30 o'clock. That is the approximate location of the famed Cerro Gordo mine, one of California's richest silver-lead producers. It enjoyed a heyday in the 1870s and produced bullion valued at about $17 million—1870 dollars, not 1990 dollars.

Inbound: Look left to the Inyos beyond Bartlett.

● In another 2 miles, 11.2 miles from Cartago, is the abandoned Bartlett chemical plant just right of the highway. It used brines pumped from wells far out on the lake floor and evaporated in the rectangular, diked areas south of the plant. Soda ash and boron were recovered from these brines. In some seasons, parts of the Owens Lake flats here are pinkish because of algae that turn red in their resting stage.

From here, on a clear day, you may be able to see buildings at Keeler, once a metropolis of 5,000 people, far across on the east shore at 3 o'clock. Keeler has been the longest enduring of all chemical operations on Owens Lake, salts being first harvested there in 1885.

Inbound: Southbound you get good views of the diked evaporation areas used by the Bartlett chemical plant. Note your odometer reading at Bartlett.

● The layers within a sequence of lake beds are like pages in a book recording history. To ascertain the early history of Owens Lake the U.S. Geological Survey drilled and cored a hole deep into Owens Lake beds, and so far has recognized events going back nearly 700,000 years.

① About 2 miles beyond Bartlett, at and beyond *"Visitor's Center 5 Miles,"* on the near floor of the valley at 1–2 o'clock are a number of curving parallel shoreline marks of historic

Photo N–2.
Bathtub ring beaches of historical Owens Lake at its north end just east of Highway 395. (Photo by Helen Z. Knudsen.)

Owens Lake (Photos N–2 and N–3). Like rings on a bathtub, these marks formed as the waterline receded following diversion of Owens River in 1913. Higher shoreline features were formed 10,000 to 15,000 years earlier when the lake was more than 200 feet deep and overflowed to the south. Remnants of these older shorelines are seen best along the east side, south of Keeler (see Photo M–10). Old shorelines are hard to recognize in the area just traversed because of erosion and deposition by powerful fan building streams from the Sierra.

Inbound: The bathtub rings are at 9 o'clock 1 mile south of the end of the divided highway, about 5.5 miles from Lone Pine.

Here the east-facing scarp of the Alabama Hills begins to rise just west of the road (see Photo N–3). When first seen, the scarp consists of bouldery fan gravels, but these soon give way northward to bedrock.

Inbound: You have had views of the scarp face of Alabama Hills from the aqueduct spillway and say goodbye here.

● As you drive north, look right across the valley to the west face of Inyo Mountains. They are composed largely of rocks unlike those making up the Sierra face. They are darker, and more heterogeneous, and in places you should be able to make out steeply inclined bedding (layering). These layered rocks are sedimentary deposits of Paleozoic age, those seen ranging roughly from 250 to possibly 500 m.y. old. Also present are some Mesozoic metavolcanic rocks, perhaps 150 to 200 m.y. old, and some still younger Mesozoic igneous intrusive rocks resembling those of the Sierra.

Inbound: These are seen as you drive south from Lone Pine.

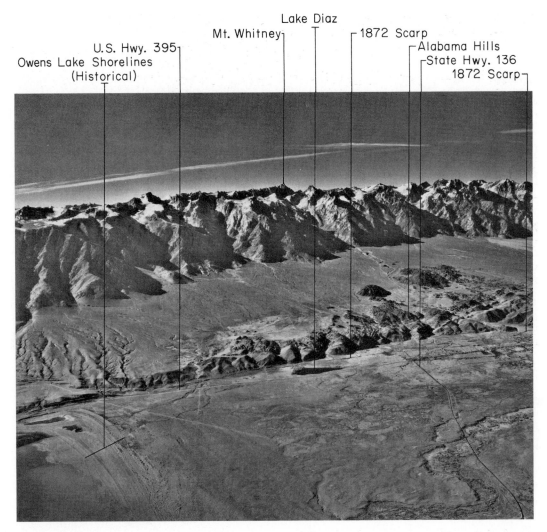

Lake Diaz
Mt. Whitney
1872 Scarp
U.S. Hwy. 395
Alabama Hills
Owens Lake Shorelines
(Historical)
State Hwy. 136
1872 Scarp

Photo N–3.
Low-altitude oblique air view of Alabama Hills and east face of Sierra Nevada near Lone Pine. (U.S. Geological Survey air photo, GS-OAI-5–17.)

● Keep an eye on the Alabama Hills. Their scarp increases in height and within 0.8 mile becomes a bedrock face, composed first of dark-brown Triassic metavolcanic rocks (200 m.y.) that give way to more massive rugged outcrops of Sierra granitic rocks back of Diaz Lake, a pond left of the highway approaching Lone Pine (Photo N–4).

Inbound: Views of the Alabama Hills are about the same southbound as northbound.

② Diaz Lake occupies a sunken strip between two fault scarps created at least in part during the Owens Valley earthquake of 1872. One forms a low, east-facing, linear bank about 10 to 15 feet high extending along the west side of the north end of Diaz Lake. It extends into the

Photo N–4.
Diaz Lake in fault graben. Alabama Hills mostly metavolcanic rock to left and granitic rock at right tip. Lone Pine Peak at range front right of center and Mt. Whitney right of Lone Pine Peak on far-distant skyline. (Photo by Helen Z. Knudsen.)

meadow to the north where trees grow on its face. The clubhouse at Mt. Whitney Golf Club sits above this scarp. The other scarp, which faces west, is hard to see from the highway.

● Mt. Whitney (14,494 feet), sharp pointed and treeless, is well viewed on the far distant skyline over the golf course, or better still, from opposite the Lone Pine Airport, where you can pull off the highway. Don't confuse it with Lone Pine Peak (see Photo 2–18), just under 13,000 feet, which looms so large on the left because of its range-front position.

Inbound: Whitney is seen nicely at 3 o'clock from the airport.

● Just 1.4 miles beyond Diaz Lake the road to Death Valley (Highway 136) takes off right. This is also the access to the Lone Pine Visitor's Center, a fine place to get books, maps, and information on the region.

Inbound: A good view of Mt. Whitney is at 3:15 o'clock from here.

● As you enter Lone Pine, it is time to consider the nature of Owens Valley. In cross section, it is a keystone shaped block dropped between two huge fault zones, one near the base of the Sierra and one along the Inyo Mountains front. Such a block is termed a *graben* by geologists.

This is an oversimplification, because you have just traveled several miles along the face of a fault scarp well out on the valley floor. The east face of Alabama Hills is just the top of a bedrock scarp nearly 10,000 feet high, which is mostly buried by over 9,000 feet of alluvium and other unconsolidated deposits. The bedrock floor of Owens Valley east of Lone Pine is 6,000 feet below sea level. This is evidence that the Owens Valley block has dropped down and not just lagged behind as the Sierra and Inyos were elevated. All this movement has occurred within the last few million years, and it continues today. The bedrock relief between the top of the Sierra and the bottom of Owens Valley is 20,000 feet. Owens Valley is a compound graben.

● Years ago, some traveling medicine man inveigled the people of Lone Pine into advertising the rocks of Alabama Hills as the oldest in the world. This is pure nonsense. The oldest rocks in the hills are Triassic metamorphics, only about 200 m.y. old. Just across the valley in the Inyos are rocks nearly 500 m.y. old; farther north in the White Mountains rocks approach a billion years in age, and east in Death Valley, rocks as old as 1.7 billion years have been radiometrically dated. Lone Pine has rich attractions; it doesn't need old rocks. Travel some of the back roads in the Alabama Hills to find that out.

Inbound: To travel such roads, take Whitney Portal Road in the center of Lone Pine or Moffat Ranch Road at north end of Alabama Hills.

● The 1872 Owens Valley earthquake was one of the largest historical quakes in California, possibly as great as the San Francisco shock of 1906. It is not known for certain, because there were no seismographs to measure it, and some people think it was of lesser magnitude. Considering the sparse habitation of the valley, the 23 to 29 deaths reported represented a significant percentage of the population. Fault scarps were formed in 1872, and you begin to see

them around Lone Pine. They continue as far north as Big Pine. The main line of breaking was along the Alabama Hills fault and its northward continuation, called the Owens Valley fault, not on the Sierra frontal fault.

● If you would like a close look at the 1872 fault scarp (Photo N–5), turn left at the stoplight in the center of Lone Pine onto the Whitney Portal Road. Proceed west 0.6 mile across the aqueduct and in about 200 feet turn right onto a two-track dirt road. Follow the main line of travel north, then left, then north again 0.2 mile to the crossing of a ditch of flowing water. Across the ditch take the road fork going right to where it turns left. Park wherever you wish. The scarp is to the left and north. Initially, it was thought that its full 20 feet height was created in 1872, but recent detailed studies suggest that only the lower 4 to 5 feet of the scarp were created then. The displacement was oblique, with perhaps a horizontal component of as much as 20 feet, accounting for the large magnitude of the earthquake.

Inbound: Turn right at the stop light in Lone Pine onto Whitney Portal Road and follow directions.

> As you leave Lone Pine, record the odometer reading opposite the south edge of Lone Pine park at the north edge of town.

③ In just 0.1 mile you can stop and get a glimpse at 8:50 o'clock of one of the more spectacular scarps heightened during the 1872 quake. It is a linear, light gray, bouldery face up to 20 feet high cutting the fan surface out from the base of Alabama Hills (Photo N–5). Rocks in the face of the hills changed back to metavolcanics near the Lone Pine airport.

Inbound: This view spot is 0.7 mile south of the 1872 graves site.

Photo N–5.
Fault scarp of 1872 and older 'quakes just west of Lone Pine.

● In another 0.6 mile, just before Pangborn Lane turnoff, is the grave site of 1872 earthquake victims. The graves are left of the highway on top of a steep rise, which is another and older fault scarp. The 1872 scarp is seen three-quarter mile southwest from the graves, below the white LP on the Alabama Hills and just above the aqueduct.

The highway runs along the base of this older scarp for at least 1.5 miles, and other segments of old scarps are seen farther left of the highway for another mile. Beyond that point closer to the hills you can see smaller scarps, formed in 1872, cutting across little alluvial fans, if the light and shadows are right. Don't take the Los Angeles Aqueduct embankment for a fault scarp.

Inbound: Watch for the south (second) turnoff onto Pangborn Lane 4 miles beyond the spillway; graves are just beyond.

● An area of trees and dead snags lies just east of the highway 2 miles north of Lone Pine park. This area was flooded during the 1872 'quake when Owens River turned into this spot, creating a shallow pond. The flooded area extends to the aqueduct spillway, 4.7 miles from Lone Pine park.

Inbound: The formerly flooded area starts near the spillway and extends for nearly 3 miles south.

● As you emerge from behind the abrupt end of Alabama Hills, the full majesty of the Sierra front appears (Photo N–6). The crest of the range towers 10,000 feet above you and 8,000 feet above the head of its alluvial apron. From here to Poverty Hills, a little south of Big Pine, this front is a spectacular example of a major fault scarp, and you have a wonderful opportunity to admire and enjoy it along that route.

Inbound: From Poverty Hills south to Alabama Hills, views of the Sierra front are incomparable.

● The aqueduct, in a large dirt ditch, is crossed 2.8 miles north of the spillway, between two curves in the highway.

Photo N–6.
Mt. Williamson and imposing Sierra Nevada front as viewed from east, between Lone Pine and Independence.

Inbound: This occurs where the highway curves left 1.7 miles beyond Manzanar relocation center road.

● Left at 9 o'clock from the abandoned site of the unforgivable Manzanar World War II relocation center, with trees, streets, and pagoda-like buildings, 9.3 miles from Lone Pine park, is Mt. Williamson (14,375).

Inbound: Southbound travelers get a superb view of this huge spectacular mountain all the way south from Poverty Hills (see Photo N–6).

● Approaching Independence, the crest of the Inyo Mountains has become somewhat lower and less rugged, because the core of the range is made up of a large granitic mass that weathers readily and uniformly. You can make out the approximate outline of this body by its lighter color and gentler terrain, compared to darker rocks into which it has been intruded.

Inbound: Southbound you see this body from a few miles north of Independence to a few miles south.

● You may have recognized that the natural course of the Owens River is indicated by a line of trees far over on the east side of Owens Valley. Old maps show that the river lay close to the base of the Inyos, not in the middle of the valley. It had been displaced eastward by huge alluvial fans from the Sierra. Sierra streams have more water and debris than the smaller intermittent streams flowing from the Inyos.

Inbound: The river line is seen best going south from Independence.

● The line of the 1872 break on the Owens Valley fault passes about 3 miles east of Independence and is marked by a prominent scarp in alluvium. To see it, turn right on Mazourka Canyon Road just at the south edge of town and drive 3.2 miles to just beyond the first distinct bend in the road. It's useful to drive 0.5 mile farther to the dry bed of the Owens River, turn around, and come back to the scarp to appreciate fully its abruptness and linearity compared to other drop offs coming down Mazourka Canyon Road. A well-graded gravel road goes north along the brink of the scarp, and the scarp face sports a good growth of bushes and trees.

Inbound: Going south from Independence watch for Mazourka Canyon Road right at the edge of town.

● At the edge of Independence is a good small Inyo County picnic area on the left opposite Mazourka Road. Proceed to the Inyo County court house in the center of town. The rock in front of the courthouse is a weathered chunk of Bishop tuff, of which more is said later.

From there a detour west for three blocks on Center Street to the small Eastern California Museum is well worthwhile. Parking is easy and admission is free. The museum has a particularly fine collection of Native American baskets. Don't overlook the outdoor displays to the west and south of pioneer farm equipment and buildings.

From the museum look east directly down Center Street, over the courthouse, to the crest of Inyo Mountains. There you should, in proper lighting and visibility, be able to make out Paiute Monument (Winnedumah), a natural solid granite tower about 80 feet high. It was in view coming into Independence and remains visible for many miles going north.

Inbound: The courthouse on the east side of the main street is easily recognized, then follow directions.

Segment O—Independence to Bishop, 41 Miles

> ▶ Record the odometer reading opposite the courthouse.

● The front of the Sierra west of Independence (Figure 3–19) is disrupted by uncharacteristic projecting spurs and forelying knobs. You tend to think of the Sierra frontal fault as a simple, single, fracture. Actually, it is a complex zone consisting of many fractures, mostly parallel. Locally fault relationships can be more complex, as here, and the topography of the mountain face is irregular. Northward the Sierra face continues to be less linear, reaches of high straight scarp are offset or separated from one another by projections or reentrants.

Inbound: The foothills along Oak Creek two to three miles north of Independence are parts of a projecting spur.

● Turn off left to Mt. Whitney fish hatchery, well worth a visit and a nice place to picnic on a hot summer's day, just over 2 miles out, shortly after the four-lane divided highway begins.

Inbound: Watch for the fish hatchery road about 0.5 mile beyond the billboards, just before the end of the four-lane divided highway.

● Paiute Monument (Winnedumah), formed by erosion of jointed granitic rock, is visible on the Inyo Mountains skyline at 3 to 3:15 o'clock driving north from Independence. It is best seen 4 miles out where the divided four-lane highway ends.

Inbound: Winnedumah is in view at 9:15 o'clock on the Inyo Mountains skyline at the start of four-lane divided highway and for most of the distance into Independence.

● Starting 2.5 miles out, and extending for 0.5 mile, is a cluster of large roadside billboards concentrated here because the highway passes through a Native American reservation where

O Nio'n Valley Rd has gigantic boulders [see Ansel Adams photo of Mt. Williamson]

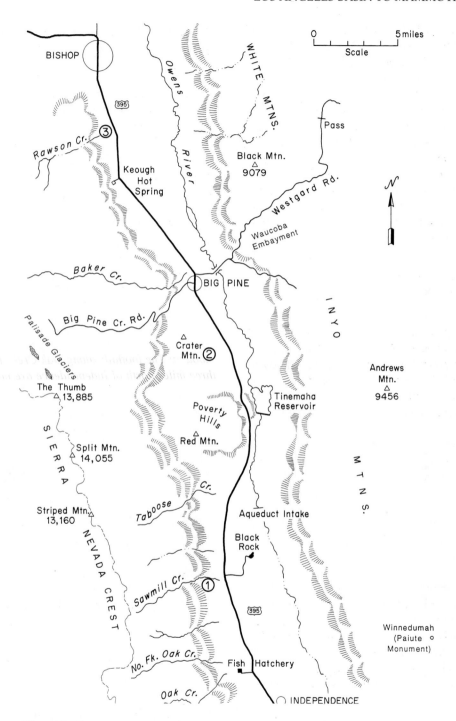

Figure 3–19.
Segment O, Independence to Bishop.

they are permitted. The paucity of billboards throughout most of Owens Valley reflects the City of Los Angeles ownership of most of the land and the city's policy prohibiting billboards.

Inbound: The billboards mar the landscape, coming or going.

● At the end of the divided highway, four miles out, on both sides of the valley at 10–11:30 and at 2 o'clock are the cinder cones and lava flows of the Taboose-Big Pine volcanic field (Photo O–1). Red Mountain at 11:30 o'clock is one of the largest cinder cones in the field. These volcanic features are not very old, as evidenced by some of the lavas that cover alluvial fans, and by cinder cones that are not greatly gullied by erosion.

The Sierra and Inyo frontal faults probably extend deep enough to provide access for molten lavas. Composition of the lavas and foreign inclusion within them suggest they came from considerable depth. Lava and cinders were spewed from 30 or more individual vents, some aligned along known faults. Eruptions occurred in at least four separate episodes all within the last 100,000 to 200,000 years.

Inbound: This volcanic field starts at Crater Mountain just south of Big Pine and extends almost to the Black Rock Springs Road, 20 miles south. Southbound views of it are good.

① Nearly 8 miles out, approaching a yellow and black sign *"Elk Next 15 Miles,"* look at the base of the Sierra carefully. At 9 o'clock is a narrow V-shaped slot cut into the steep mountain face by Sawmill Creek. If the light is right you will see masses of black rock adhering to both walls of this slot a little above its mouth. These are remnants of lava that flowed down the canyon, filling it to a depth of 150 feet. Stream erosion has since cut a narrow cleft through the lava. These flows are interesting to geologists because they have been dated as roughly 60,000 years old by a special potassium-argon method. Far-

ther up the canyon a glacial moraine rests on top of the lava, showing that this stage of glaciation occurred less than 60,000 years ago. Still farther up Sawmill Creek the same lava unit rests on top of older morainal deposits that, therefore, have an age greater than 60,000 years.

Also, at 10:30–11 o'clock, from the elk sign, one sees five or six cinder cones close to the base of the mountains and extensive lavas extruded from these vents.

Inbound: Sawmill Canyon and its lava are seen well 0.3 to 0.4 mile beyond Black Rock Springs Road.

●Just 0.4 mile farther is Black Rock Springs Road, going both east and west. If you are interested in fish, follow the east branch 1 mile to the Black Rock rearing ponds that house hundreds of trout, from fingerlings to 13 inches long. They are fun to see and visitors are welcome. Note the total inclosure, above as well as on the sides, of the ponds to protect against fish-eating birds.

Inbound: Black Rock Springs Road is 0.4 mile beyond the edge of the Taboose lavas, just over two miles from the last rest area.

● In 0.3 mile from Black Rock Springs Road, Highway 395 passes onto some lava flows, and you get a closer look at their ragged surface. If you stop driving, you will see that the lavas are twisted, jagged, blocky, and generally full of small holes formed by bubbles of gas trapped in the solidifying rock. These are what geologists call *basaltic aa lavas,* relatively low in silica and high in iron and magnesium.

● Beyond *"Rest Area 1 Mile"* is a good view of Red Mountain cinder cone (see Photo O–1) at 11 o'clock. White Mountain Peak (14,248 feet) is far away at 11:55.

Inbound: Red Mountain is in good view at 3:30 o'clock about 2.2 miles beyond the end of the four-lane highway over Poverty Hills.

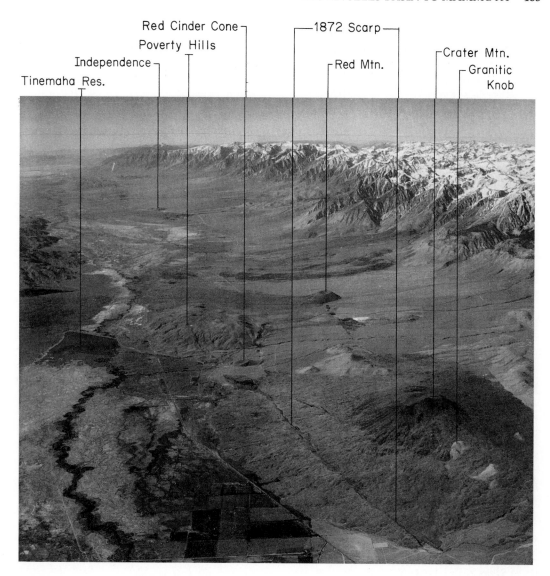

Photo O–1.
Air view south down Owens Valley from over Big Pine. (Photo by Roland von Huene.)

Photo O–2.
Segment of old concrete highway, eight feet wide with turnouts. In the early 1920s the only pavement between Bishop and Mojave. Locale is south side of Poverty Hills, just west of Highway 395. (Photo by Helen Z. Knudsen.)

● A special feature, for inbound travelers mostly, is to look close along the west side of the highway, 1.4 miles beyond the Poverty Hills four-lane highway, for small patches of narrow concrete pavement (Photo O–2). Highway repair materials (gravel) are piled on some of them. They are remnants of an 8 mile, 8 foot wide concrete highway, with turnouts, that in the 1920s was the only pavement between Mojave and Bishop.

● A mile north of Goodale Creek Road at about 11 o'clock and up-slope from Red Hill is a light-colored pumice quarry and processing plant, an unusual feature in the midst of dark basaltic volcanics.

Inbound: This pumice operation is best seen southbound, beyond Fish Springs hatchery road by looking up-slope over the top of the little red cinder cone.

● Beyond the Aberdeen Station-Taboose Creek roads, at "4 Lanes 2 Miles Ahead," look right at 2:45–3 o'clock to see two cinder cones and much lava at the base of Inyo Mountains. Cinders are often red, because the iron in them becomes oxidized while they are still hot and flying through the air. Volcanic gases may aid the alteration.

Inbound: These Inyo cones and lavas are in view coming down the south side of Poverty Hills.

● The eastward projecting spur constituting Poverty Hills is crossed by a four-lane highway starting about 6.5 miles beyond the rest area. The hills are composed of old metamorphosed shale and limestone (marble) intruded by granitic rock.

Inbound: You start across Poverty Hills on four-lane highway just beyond Tinemaha Creek Road.

● One mile beyond the end of the four-lane highway across Poverty Hills, look at 9:15 o'clock for a good view of a neat red cinder cone (Photo O–3) on the far side of the green (in summer) fields. To the north and a bit higher up the slope at 9:30–10 o'clock is a fault scarp in lavas aligned along the 1872 break.

Inbound: At 1 o'clock at the Fish Springs hatchery road you look right into the bowl of the red cinder cone's crater.

Photo O–3.
Small red cinder cone behind Fish Springs west of Highway 395. (Photo by Helen Z. Knudsen.)

② In another mile at about 10:30 o'clock, the large mass of nearby Crater Mountain (see Photo O–1) is dominant; keep an eye on it. Crater Mountain is a volcano perched on a granitic ridge with lavas draped over its flanks, like chocolate on a sundae. A bit farther north are two knobs of light-colored granitic rock peeking through the lavas (see Photos O–1 and O–4). These bear the delightful Hawaiian term, *kipukas,* islands in the midst of a sea of lava. If the sun is in the west, you should see several shadowed fault scarps extending northward across the lower slope of Crater Mountain from 10 to 11 o'clock. The freshest and most continuous was formed during the 1872 earthquake (Photo O–1).

Inbound: Just beyond "4000 Feet Elevation" south of Big Pine, the granitic kipukas of Crater Mountain are at 2:45 and 3 o'clock (Photo O–4). In another three miles is the fresh fault scarp in the lavas on the mountain's slope.

● Approaching Big Pine you get some excellent views, in clear weather with late afternoon lighting, of the variegated rocks composing the Inyo Mountains front to the right. Most of the sedimentary rocks exposed there are of Cambrian age, 500 to nearly 600 m.y. old.

Inbound: You can see the rocks to the left south from Big Pine.

Photo O–4.
Kipuka (island) of granitic rock surrounded by younger lava flows on north ridge of Crater Mountain (compare Photo O–1.) (Photo by Helen Z. Knudsen.)

● At the road going left to the Inyo County waste station and Big Pine cemetery, where the four-lane highway begins, look to the left about 0.5 mile to see the 1872 fault scarp.

Inbound: As you come south out of Big Pine, look right where the four-lane highway ends to see this scarp.

● Big Pine Canyon, west of town, is hard to identify because its mouth is plugged by massive glacial moraines. If you drive up Big Pine Creek, the road switches back on the inner face of a moraine just after the second creek crossing, and your subsequent route nearly all the way to Glacier Lodge is upon the lateral-moraines of the Big Pine Glacier. Glaciers reached the east base of the Sierra at many canyons north from here but at only a few to the south, the southern-most being Independence Creek.

● Where the highway curves into Main Street at the south edge of Big Pine at 8:45 on the far Sierra skyline is a dark jagged ridge with one large dominant peak. At its base is a long narrow band of ice and snow, the Palisades Glacier (Photo O–5), one of the largest and most southerly glaciers in the Sierra. It is also in view at *"School, Speed Limit 25,"* and at the Bartell/Blake roads crossing. It is still in view 0.5 mile north of town at 9 o'clock near the Westgard Pass (Highway 168) intersection.

Owens Lake — ┌Big Pine

Dissected Fans

Palisade Glacier

Waucoba
Embayment

Coyote Warp

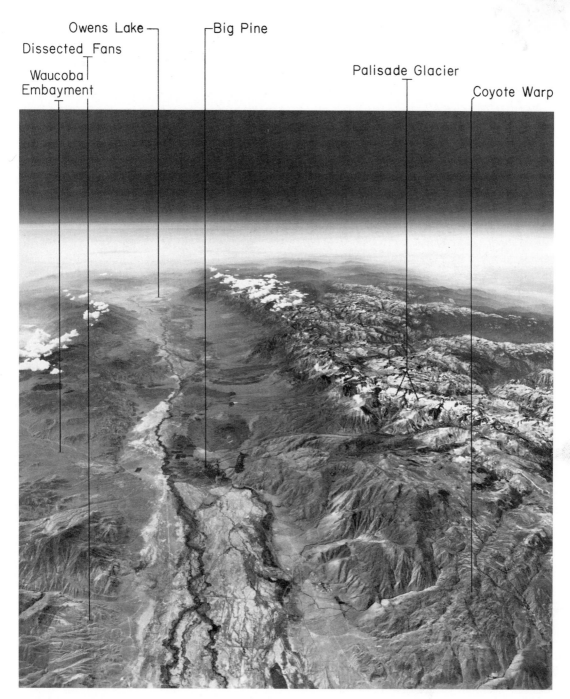

Photo O–5.
High-altitude oblique view south down Owens Valley from over Bishop. (U.S. Air Force photo
taken for U.S. Geological Survey, 018L-057.)

Inbound: You get a good chance of seeing Palisades Glacier at about 2:50 o'clock along the straight stretch into Big Pine from the Westgard junction. The view just beyond the Bartell/Blake roads at the south edge of town is better than outbound.

Existing Sierra glaciers may not be the shrunken remnants of Ice-Age glaciers. Instead, they could have formed within the last 3,000 to 4,000 years. Many lines of evidence show that the climate several thousand years ago was warmer and drier than now. Owens Lake may have dried up then, suggesting that the Sierra were denuded of perennial ice and snow. About 4,000 years ago the climate became wetter and cooler, and perennial snow banks formed at protected spots in the Sierra. Eventually they grew into new glaciers. Judging from the moraines they have built, at least some of these new glaciers attained a maximum length of 0.5 mile between 1750 and 1890 A.D. Since then they have been shrinking mostly.

▶ | Record your odometer reading in the center of Big Pine opposite the Chevron station.

● About 0.5 mile north of town, Highway 168 leads east to Westgard Pass, Deep Springs Valley, and the northern end of Death Valley. The subdued Inyo Mountains front and the lower skyline to the east are known as the Waucoba embayment (Photo O–6), which separates the Inyo and White Mountains. The white patches on the lower slopes of the embayment are uplifted and dissected lake beds, at most 2 to 3 m.y. old. Similar materials probably underlie the alluvium of Owens Valley. The unconsolidated fill in the valley here is nearly 8,000 feet thick. The greatest depth to bedrock is close to the base of the Inyo and White mountains, suggesting eastward tilting of the valley floor.

Inbound: Waucoba embayment and its lake beds are seen pretty well at 10:30 from the end of the separated highway lanes opposite the radio-telescope dishes.

● Beyond Reynolds Road, 1.7 miles from Big Pine, on the valley floor at 1:30 are the "ears" of the Caltech Radio Astronomy facility. In 1993, they consisted of two large, 130-foot diameter, two 90-foot, and numerous 33-foot parabolic antennae (dishes), mounted on rails to permit shifting for "stereo" listening.

Inbound: The radio-telescope dishes are best seen at 9–9:15 o'clock within the speedometer check zone, 2 miles beyond Keough Hot Springs.

● About six to seven miles north of Big Pine, views of the White Mountains to the right are good, especially in afternoon light. Along the base of the range are deeply dissected alluvial fans extending far up the canyons. These uplifted and dissected deposits are another expression of recent geological deformation in this area. Relationships are particularly clear ahead near the outdoor theater.

Inbound: Keep watch on the lower slopes of White Mountains from Gerkin Road to south of Collins Road to see these uplifted, dissected fan deposits.

● Most of the rocks in the west face of White Mountains from here to Bishop are Cambrian sedimentary beds (505 to 570 m.y.). The east side and crest of the range to the north are composed of older, presumably late Precambrian sedimentary formations. Locally, mostly to the east and north, are much younger (100 to 150 m.y.) granitic intrusive bodies. White Mountain Peak (14,248 feet) has been prominent on the far skyline at about 1 o'clock all the way from Big Pine.

● At Collins Road, north of Keough's Hot Spring by 0.5 mile, where the four-lane highway starts, the fan surface at 10–11 o'clock has

Photo O–6.
Waucoba embayment between Inyo and White mountains east of Big Pine. (Photo by Helen Z. Knudsen.)

a peculiar hilly appearance. You are looking at the back side of a little fault block with a scarp facing west toward the Sierra. Normally you would expect scarps associated with uplifted mountains to face the valley. Mountain-facing scarps are probably created by small-scale settling adjustments within the valley block.

③ If you'd like to escape some traffic and see a little more of this fault feature, turn left on Collins Road for 0.7 mile to Gerkin Road and turn right. Now plainly visible just to the right behind the power line towers is the scarp. It dies out northward in less than 0.5 mile. Follow Gerkin Road to rejoin Highway 395 a short distance ahead.

Inbound: Southbound travelers can make this loop by turning off on Gerkin Road just short of the outdoor theater and returning to Highway 395 on Collins Road.

● The front of the Sierra from Big Pine north is less abrupt and imposing than farther south (see Photo O–5). This part of the Sierra is regarded as having been warped more than faulted. As you will see shortly, it constitutes a spur north of which the main mass of the

Sierra is offset westward about 10 miles. This structure, called the Coyote Warp, bends down under Owens Valley to the east and plunges north under Bishop.

Inbound: You have seen the blunt plunging nose of Coyote Warp driving into Bishop on Highway 395.

● The Wheeler Crest part of the Sierra front seen at 9–11 o'clock from Gerkin Road is awesome by comparison to the Coyote Warp.

Inbound: You have already seen Wheeler Crest close up and the contrast is striking.

Segment P—Bishop to Mammoth, 40 Miles

● Highway 395 to Mammoth (Figure 3–20) turns left at the Y at the north end of Main Street in Bishop.

▶ Note your odometer mileage here.

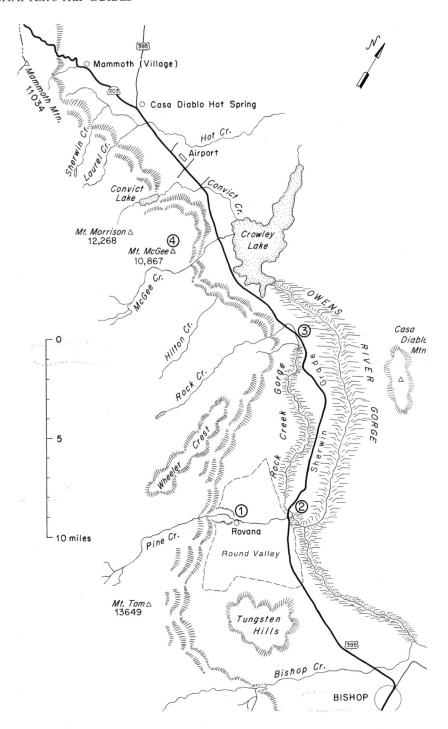

Figure 3–20.
Segment P, Bishop to Mammoth.

Ahead you see the 10 mile westward offset of the Sierra front beyond the north end of Coyote Warp.

● West of Coyote Warp is Bishop Creek country. The unusual northward course of Bishop Creek, within the mountains, may have been determined by the relative uplift of the Coyote Warp. West of the mouth of Bishop Creek are subdued, brush covered, gray hills through which the Bishop Creek highway now winds. These hills are composed of bouldery materials laid down largely by early advances of the Bishop Creek glacier. They remain in view for the next four miles.

Inbound: You see the Bishop Creek setup well from the gun club ranges stop.

● Roughly 2.5 miles west of the Y and 1 mile beyond Barlow Lane, a small turnout on the right at the west edge of the trees makes a good place to stop. From here about halfway up the face of Coyote Warp at 9 o'clock, one sees a smooth bench sloping gently northward. It is capped by a dark lava flow resting on paler granitic rocks. If the light is good and your eyes are sharp, or you have binoculars, you may be able to make out some large, light-colored boulders resting on the lavas about where the trees begin. These granitic boulders were emplaced by a glacier flowing out of higher country to the south early in the Ice Age.

Inbound: These features are in view from a small right-side turnout 0.1 mile beyond the Bishop gun club ranges.

● At 3.3 miles from the Y just beyond the Bishop gun club ranges is a good turnout parking place on the right, worth a stop. From here the skyline at 11:45 o'clock is dominated by Mt. Tom (13,648 feet). Massive irregular rocky embankments at the mouths of cirques on the east and northeast sides of Mt. Tom are moraines, behind which may be rock glaciers. A *rock glacier* is an accumulation of large angular boulders that creeps or has crept slowly

down slope. The lower and nearer knobby, rocky terrain is Tungsten Hills, named for its many tungsten mines and prospects.

In the foreground at 10 o'clock is a dark volcanic cinder cone. The mouth of the gorge of lower Bishop Creek is also at 10 o'clock. The narrow, gray curvilinear ridge to the right is a glacial moraine. The midway bench with glacial boulders at tree line on the Coyote Warp is at 9:10 o'clock.

Inbound: This cone is well seen at 3 o'clock 0.6 mile beyond Ed Powers Road, and the Mt. Tom moraines and rock glaciers have been visible in earlier views coming south. They can be seen again from the small right-side turnout just beyond the Bishop gun club ranges right of the highway.

● At 0.5 mile beyond Ed Powers Road a gentle descent begins. Round Valley, ahead, occupies a down dropped block lying between the east-facing Sierra fault scarp and this west-facing rise you are descending, which was formed by a fault and a warp. It parallels our route for the next 5 miles.

Inbound: You ascend this gentle rise before getting to Ed Powers Road.

● In less than a mile, a little beyond the Forest Conservation camp sign, Pleasant Valley Dam road turns off right down a wide valley that formerly carried a large drainage out of Round Valley. Within the next 2.5 miles the highway gradually converges to an intersection with this old stream course (Photo P–1), at the point of rocks and Mill Creek Road ahead. There, the cliffs east of the old stream valley are composed of Bishop tuff, of which you see a lot going up Sherwin Grade ahead.

Inbound: You look directly down this wide valley beyond Mill Creek Road.

① Near "*Pine Creek Road, Rovana 3, Lower Rock Creek,*" is a good place to stop. Across the valley at 10 o'clock are the morainal ridges protruding from the canyon of Pine Creek (Photo P–2). The settlement at the mouth of the

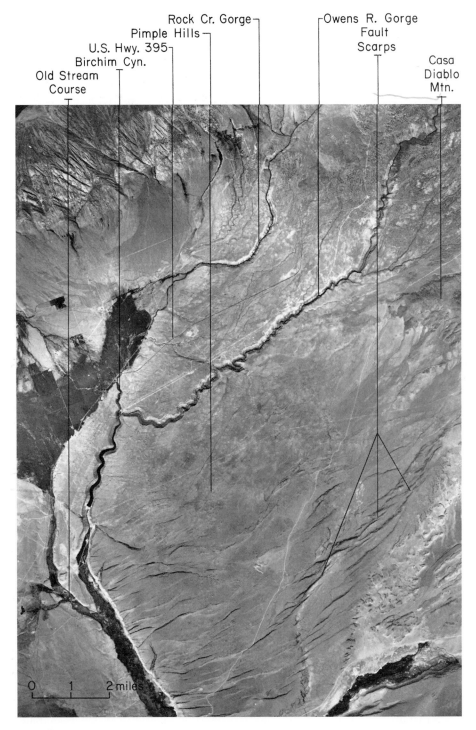

Rock Cr. Gorge

Pimple Hills

Owens R. Gorge
Fault
Scarps

U.S. Hwy. 395

Birchim Cyn.

Casa
Diablo
Mtn.

Old Stream
Course

0   1   2 miles

Photo P–1.
High-altitude vertical view of Sherwin Grade area, north to right, scale in lower left.
(U.S Air Force photo taken for U.S. Geological Survey, 018V-058.)

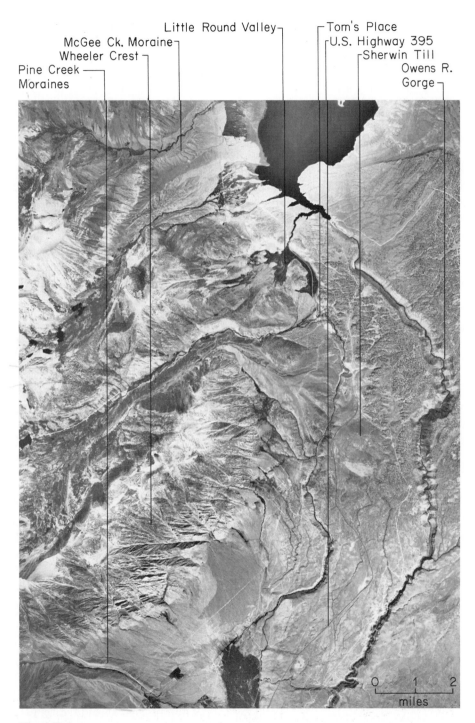

Photo P–2.
High-altitude vertical view of upper Sherwin Grade-Crowley Lake area, north to upper right corner, scale in lower right. (U.S. Air Force photo taken for U.S. Geological Survey 018V-059.)

creek is Rovana. Farther up Pine Creek one of the largest tungsten-processing mills in the country once operated. The mines supplying the mill were higher, 9,500 to 12,000 feet, and farther back in steep, glaciated terrain near the headwaters of a north branch of Pine Creek. They cannot be seen from the highway. The tungsten ore was from what geologists call a "contact deposit." In this instance the contact was between old metamorphic rocks, which were once partly limestones, and younger intrusive granitic rocks. Although the ore is in the altered rocks, the principal ore elements, tungsten and molybdenum, were formed by hot fluids and vapors given off by the intruding igneous body. This was long one of the richest tungsten operations in the United States.

Inbound: Pine Creek, Rovana, and the Pine Creek moraines are seen between the Rovana and Mill Creek turnoffs.

● The high, abrupt Sierra face west of Round Valley is Wheeler Crest. It is composed of massive granitic rocks with steep jointing that produces castellated cliffs when eroded. Keep an eye on this face as you go north. (Read ahead in the guide.)

Inbound: The views of Wheeler Crest as you come down Sherwin Grade and from Round Valley are great.

② Just before the highway swings slightly right to start an ascent of Sherwin Grade, a quick look right reveals the gorge of Birchim Canyon (see Photo P–1) through which Pine and Rock creeks now flow out of Round Valley to Owens River. Earlier they presumably followed the old stream valley seen five miles back.

▶ | Note your odometer at the intersection with Gorge Road and Lower Rock Creek Road.

Inbound: Stream diversions like the Pine and Rock are more common than generally realized.

● The ascent of Sherwin Grade takes place on top of the Bishop tuff. In Long Valley, 20 miles to the north, about 750,000 years ago, a huge volcanic explosion ejected clouds of hot glowing ash and rock particles that spread over the surrounding landscape for great distances. These clouds, made dense by suspended particles, flowed rapidly by gravity and were kept mobile by gases released from hot volcanic fragments. When the material came to rest, it was still so hot that the particles fused and partly recrystallized, creating a coherent rock. This phenomenon was like the "glowing cloud" that wiped out the city of Saint Pierre and 30,000 inhabitants in 1903, when the Mt. Pelee volcano erupted on the West Indies island of Martinique.

Inbound: If a new caldera were to form in or near Long Valley, which is not impossible, the resulting glowing cloud would create havoc.

When the Long Valley eruption ceased, a nearly level sheet of welded ash at least 500 feet thick flooded and largely submerged features of the preexisting landscape. Only large features, such as Casa Diablo Mountain, stick through as islands in the sea of tuff. The sheet extended from Mono Lake 65 miles to south of Big Pine. One hundred and fifty cubic miles of volcanic debris were involved—enough to cover all of Los Angeles County 200 feet deep (a sobering thought). Subsequently, the sheet has been faulted (see Photo P–1) and warped, eroded, and buried. Only remnants are preserved today, and you are traversing the largest. The gentle incline you ascend is the result of tilting of the sheet.

Inbound: The best view of Casa Diablo Mountain is at 9:15 o'clock from the Scenic Point.

You can see what Bishop tuff looks like as you pass through road cuts ascending Sherwin Grade. Fresh exposures are mostly pink, but the tuff weathers brown. The white zone near the top of many cuts is a result of near-surface

Photo P–3.
Columnar joint rosette around fumarole pipe in Bishop
tuff, as exposed in east wall of Owens River gorge. (Photo
by Helen Z. Knudsen.)

weathering that causes deposition of calcium
carbonate along fractures. Close inspection
shows the rock to contain fragments of pumice,
shiny particles of feldspar and quartz, much
dull, nondescript ashy material, and small angular chunks of rock, mostly volcanic.

People particularly interested in the Bishop tuff
can see good exposures in the walls of Owens
River gorge by turning off right on Gorge Road
and walking down side roads off it into the
gorge (Photo P–3).

*Inbound: Most cuts show that the upper few
feet of the Bishop tuff has been browned by
weathering. The white band extending to the
surface is misleading, because the white is limited largely to cracks, the whole rock is not
white. To see the details of the tuff's makeup,*

*you have to stop and pick up two or three
pieces. Examination with a hand lens or any
magnifying glass helps.*

● As the highway rises start looking left and
right. You will see many small pimple-like
hills, often in clusters (Photo P–4). They represent spots where jets of gas from the ash found
their way to the surface, hardening the tuff en
route. Erosion has converted these firmer spots
into little conical hills, 20 to 75 feet high. They
are called *fumarole pimples* and are especially
abundant at 9 and 3 o'clock from *"Tom's Place
9 Miles."*

*Inbound: The fumarole pimples seen at the
scenic point should help you recognize similar
forms going down Sherwin grade, particularly
left of the 6,000 feet sign. The other lanes of the
highway cut through one there.*

Photo P–4.
Fumarole pimples on surface of Bishop tuff alongside Gorge Road. (Photo by Helen Z. Knudsen.)

● About 2.3 miles farther the highway enters a long straight stretch. Ascending it, you see Casa Diablo Mountain about 5 miles away at 1 o'clock. This particular "House of the Devil" is an island or *kipuka* of older granitic rock rising through the Bishop tuff. On the far skyline at 2 o'clock is White Mountain Peak (14,248), composed of old metamorphosed volcanic rocks intruded by younger granitic bodies. A cluster of fumarole pimples is seen on the tuff surface in the mid-distance at 9–11 o'clock.

Inbound: About 2 miles beyond the start of divided highway is a scenic turnout worth a stop. In the foreground is a field of fumarole pimples. Distant features viewed include Mt. Tom, Round Valley, Pine Creek moraines, Wheeler Crest, Tungsten Hills, Bishop Creek Country, Coyote Warp, Bishop, Owens Valley, Waucoba Embayment, Inyo Range, White Mountains, White Mountain Peak (11 o'clock), Casa Diablo (9:15), and the Owens River penstock.

You are here traveling a course roughly parallel to and about midway between the gorges of Rock Creek, to the left, and Owens River to the right (see Photos P–1 and P–2). These streams have cut steep-walled gorges hundreds of feet deep into the Bishop tuff and locally down into the underlying granitic rocks.

Inbound: Neither inbound nor outbound do you see these gorges, but they are accessible via the Gorge Road and old Rock Creek Road, respectively.

● The highway eventually curves left, and about where the piñon trees begin you get a good view at 8–9 o'clock toward Round Valley. You can now see better the complex of morainal ridges protruding from Pine Creek canyon. Big yellow pine trees appear about a mile beyond the *"Inyo National Forest"* boundary.

Inbound: Southbound travelers get a fine view of Mt. Tom as they come down Sherwin Grade, and also of the Pine Creek moraines and Round Valley.

● Just 1.2 miles beyond the Inyo Forest boundary you pass from Bishop tuff onto granitic rocks, exposed in a double-walled road cut. In another 1.3 miles the highway cuts through good bouldery glacial *till*, appropriately named the Sherwin till. Till is unsorted rock debris deposited directly from glacial ice. The best exposure is 8.5 miles from Gorge Road.

Inbound: This exposure is just beyond the "Brake Check Area" sign. The highway starts passing through cuts in weathered granitic rock just 0.4 mile beyond the brake check parking area.

● At *"Sherwin Summit, Elevation 7,000,"* cliffs of Bishop tuff are at 1–3 o'clock. The tuff overlies the Sherwin till composing subdued brush covered slopes right of the highway, south of the cliffs.

Inbound: The road to Owens River gorge parallels these cliffs.

● White material exposed in the second road cut, 0.5 mile beyond the summit, is air-fall and flow pumice at the base of the Bishop tuff.

Inbound: This is the next cut beyond the big cut with pumice and till. Dead ahead, the rounded skyline hill with relay equipment on top is all Sherwin till. **Special Note:** *Inbound travelers wishing to see the Bishop tuff exposures in Owens River gorge can turn off shortly to the east on a dirt road that will lead to the gorge and along it to a junction with Gorge Road that takes them back to 395 near the base of Sherwin Grade. The turnoff is not signed nor easy to see. It is just short of the "Sherwin Summit, Elevation 7,000," sign, 1.1 miles south of the big pumice cut. If the turnoff is overrun, turn around at Sherwin Summit, the side road is easier to spot northbound.*

③ One mile from the summit and just beyond *"Lower Rock Creek, Swail Meadow,"* you descend toward Rock Creek and pass by a deep right-side road cut exposing important relationships. The white material is the basal Bishop pumice, and toward the north end of the cut it overlies a coarse bouldery deposit, which is Sherwin till (Photo P–5). Minerals in the pumice have been dated by the potassium-argon method as 750,000 years old. This fixes the age of the Sherwin glaciation at about 800,000 years, because the till was extensively weathered and eroded before the pumice was deposited.

Inbound: This big cut is just a whoop and a holler beyond the crossing of Rock Creek and has a huge flat turnout area on the right.

● About 1 mile past Lower Rock Creek turnoff, where the four-lane highway becomes divided, look at 9:30 o'clock to see the mouth of the upper Rock Creek gorge. The glacier that deposited Sherwin till came from Rock Creek, but it did not come through that narrow defile, which is post-Sherwin in age. Instead, it followed a course along Whiskey Canyon about 1.5 miles south.

Inbound: It's difficult to see the mouth of upper Rock Creek gorge southbound.

● Rock Creek makes a sharp turn to the south just right of the Crowley Lake Drive exit at Tom's Place (0.6 mile ahead) to flow through its long lower gorge to Round Valley. Before diversion to that course long ago, it continued northeastward through Little Round Valley about parallel to the present highway into Owens River at the Crowley Lake dam site.

Inbound: In this direction you are driving up the abandoned course of old Rock Creek.

● In a mile beyond the Crowley Lake Drive exit (at Tom's Place) note the contrast between jointed, gray granitic rock in the large knob left of the highway and massive, tan Bishop tuff in cliffs to the right. The ditch just left of the road, more visible southbound, is used at times of high water to divert part of Rock Creek along its old route into Crowley Lake, so the water can be run through powerhouses and have all its kilowatts extracted. The narrows at the north corner of Little Round Valley, harboring an arm of Crowley Lake, mark the old abandoned Rock Creek course.

● Road cuts in the hills beyond Little Round Valley are in Bishop tuff.

Inbound: Exposures of Bishop tuff with cavernous weathering are good in road cuts beyond the Crowley-Hilton exit. Crossing Little Round Valley, Bishop tuff cliffs to the left and eventually the big knob of jointed granite to the right are worth watching for.

Photo P–5.
Big Pumice cut on Highway 395 near Rock Creek crossing. Vertical structures in white Bishop pumice are clastic dikes. Bouldery debris low on left is buried Sherwin till. Pumice is 730,000 years old.
(Photo by Helen Z. Knudsen.) *(from Long Lake Volcano)*

● Just beyond the underpass at the Crowley Lake-Hilton Creek exit, look to the mountain front at 9 o'clock to see the massive Hilton Creek moraine, a huge embankment of bouldery debris. This could be termed a dump moraine, because Hilton Creek glacier came to the lip of its hanging valley and just dumped its load of debris down over the steep mountain face below. The bouldery finger-like ridge projecting from the base of this moraine, one mile beyond the Crowley Lake overpass, was probably made by a debris flow off the moraine.

Inbound: These features are well seen from the scenic turnout.

● At the photo point turnout, 0.7 mile beyond Crowley Lake exit, at 10–11 o'clock on the far side of the brush-covered flat, is the spectacular right-bank, lateral-moraine complex of McGee Creek, fully 700 feet high (Photo P–6). (Banks of streams and canyons are designated right or left facing downstream.) This moraine was built during at least two glacial stages, as indicated by the shadow-enhanced gullies on its outer slope that terminate well below its crest. This is also a good place to observe the bouldery debris-flow finger projecting from the base of the Hilton Creek moraine.

Inbound: This impressive McGee Creek composite lateral moraine is hard to appreciate southbound unless you stop so you can look back to it. A good place is the north end of the scenic point turnout a little short of 2 miles beyond McGee Creek road. From here you also can see the Hilton Creek moraine and its projecting debris-flow finger nicely.

From Hilton Creek to the Mammoth turnoff the Sierra Nevada rocks have many shades of brown, red, white, and black. These variegated units are part of a large metamorphic septum. Originally they were shales, sandstones, and limestones. Some contain identifiable marine fossils showing that the oldest layers were deposited about 400 to 500 m.y. ago.

Photo P–6.
Large right lateral moraine deposited by McGee Creek glacier. Gullies terminating short of crest indicate at least two stages in moraine construction.

Inbound: The same holds inbound but is best recognized by looking at stream gravels from the mountains.

● Crowley Lake is artificial, but about 75,000 years ago a large, glacier fed, natural lake at least 200 feet deep lay in this basin, covering an area of 90 square miles. This lake's outflow helped deepen the Owens River gorge. Crowley Lake occupies only the south part of Long Valley, which is the first of several depressed basins along the foot of the northern Sierra. Mono and Bridgeport are the two basins lying next north. These are geologically recent structures deeply filled with unconsolidated deposits. The fault cutting the moraines at McGee Creek (Photo P–7) is just one expression of geologically recent deformation in the Long Valley area.

Inbound: Unfortunately, the McGee Creek faulted moraines are hard to see unless you drive up McGee Creek, but Crowley Lake is all yours.

● Long Valley, as well as the hilly plateau immediately north and the embayment of the Sierra front at Mammoth, are part of a huge, 10 by 20 mile, elliptical structure, a *caldera,* formed by collapse following the great volcanic explosion and expulsion of ash that made the Bishop tuff, about 750,000 years ago. The part of the ash thrown high into the air is widely distributed in Pleistocene sedimentary accumulations over parts of southern California (especially within desert lake beds) and as far east as Kansas and possibly Washington, D.C. Similar explosions created three separate caldera in Yellowstone Park, but there, as here, the calderas have subsequently been filled by younger lavas, so they no longer form a depression in the landscape like that at Crater Lake, Oregon.

Starting in the 1960s and extending at least to the early 1990s, a series of earthquakes has shaken this region. Tiltmeters, leveling surveys, and seismic records showed that the land

Photo P–7.
Fault scarp cutting moraines at mouth of McGee Creek Canyon. Compound nature of right lateral moraine (on left) is indicated by successive ridges and terminated gullies. (Air photo by Roland von Huene.)

under the center of the Long Valley caldera was swelling. One aspect of the seismic records, called *harmonic tremors,* suggests that molten magma was stirring around at a depth of a mile or two. As required by law, the Director of the U.S. Geological Survey issued a volcanic hazard warning, much to the displeasure of local residents, because of the resulting drop in real estate values and the reduction in numbers of skiers and vacationers.

Deposits of pumice show that volcanic activity has occurred about once every 200 to 300 years in the Inyo and Mono domes and craters. It has been roughly 250 years since the last eruption. No volcanic material has yet been erupted in the 1900s, however, and the current subsurface activity may taper off without causing a surface explosion.

The geological record suggests that eventually an eruption is almost certain to occur. It could well be small and local, like the events that formed the Inyo domes. The ash from such an eruption could be a local nuisance and dangerous, but the countryside is not likely to be buried in lava. An explosion of the size that created the Long Valley caldera would be a major disaster for a large part of southern California, easily a thousand times more destructive than the Mt. St. Helens eruption of 1980; stay tuned.

Inbound: The rock cliffs left of 395 northeast from Convict Lake turn-in are composed of rhyolitic lava, part of the filling of Long Valley caldera. The caldera is entirely obscured by Bishop tuff and younger rhyolitic and basaltic lavas. You can get only a hazy impression of its size and form from the present Long Valley, the Mammoth embayment and Glass Mountain on its far eastern rim.

● Crossing McGee Creek, look to the far eastern skyline at 2:45 o'clock to see Glass Moun-

tain, so named because the rocks composing it contain abundant volcanic glass (obsidian). In a mile beyond the big highway maintenance station, Mammoth embayment and Mountain (11:30) come into view, backed on a clear day on the far 11:40 skyline by the Minarets and Ritter and Banner peaks.

● After rounding the corner and crossing Convict Creek, the massive morainal system of Convict Creek comes into view about a mile to the left. Study Photo P–8 to get a sense of relationships. The road to Convict Lake, one mile ahead, takes off southwest and crosses a large lobate morainal complex formed by the last glacial advance. At an earlier stage Convict glacier followed a more southerly route along the base of McGee Mountain and built the huge lateral moraine, nearly 1,000 feet high, which you see by looking back at 8 o'clock from the Convict Lake turnoff.

Inbound: Southbound travelers get a better and easier look at these Convict moraines opposite and beyond Mammoth airport. A stop at the Convict Lake turn-in gives good views of the bouldery young moraine complex crossed by the Convict road and the smoother but huge old lateral moraine (see Photo P–8).

● Beyond Convict turnoff, you get a good look at lone-standing Mammoth Mountain at 11 o'clock on the skyline (Photo P–9). If it's winter and everything is normal, the mountain will probably be crowned by its own local cloud.

Inbound: Mammoth Mountain was selected as a ski area because it gets more than normal snow from this and other clouds.

Incidentally, this is *True Grit* country. Those of you with geological knowledge might have recognized outcrops of the Bishop tuff in parts of that John Wayne movie, as well as the scenes along Hot Creek. You should also have spotted

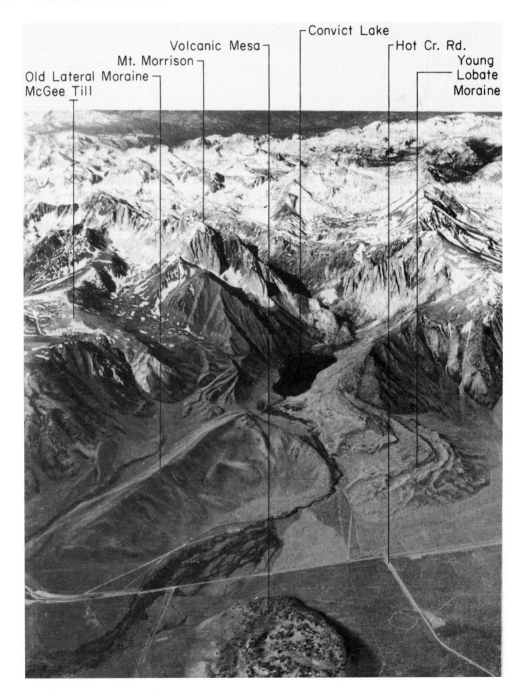

Photo P–8.
Looking south to Convict Lake over Convict Creek glacial moraines. (Air photo by Roland von Huene.)

Photo P–9.
Free-standing Mammoth Mountain, a possibly not yet extinct volcanic construct on rim of Long Valley caldera. (Photo by Helen Z. Knudsen.)

Photo P–10.
Mount Morrison, with snow-mantled Convict glacier's youngest moraine in foreground. (Photo by Helen Z. Knudsen.)

Mt. Morrison (Photo P–10) in the background from time to time. It is the high pointed skyline peak at 8:45 o'clock, made up of steeply inclined layers of metamorphic rock. Much of the film was made in Colorado, so don't look for the large groves of gorgeous quaking aspen.

Inbound: Mt. Morrison (Photo P–10) is the high, sharp peak at about 2:30 o'clock left of Convict Lake. Don't confuse it with the closer Laurel Mountain at/ 3 o'clock.

Photo P–11.
Foreground boulders and bouldery deposit on background ridge are remnants of the old McGee morainal deposit on top of McGee Mountain.

④ Beyond Convict Lake turnoff, a look back to about 8 o'clock on the high skyline (2 o'clock for southbound travelers) should show you a deposit of large, light-colored boulders resting on darker rocks about where the trees are thickest, provided everything isn't covered by snow. This is the McGee till (Photo P–11), one of the oldest glacial deposits of the Sierra. The boulders are granitic (Photo P–11), some up to 20 feet across, which were carried from upper McGee Creek and deposited on the crest of this ridge by the McGee Creek glacier long, long before the moraines in the present canyon were laid down.

Inbound: Southbound travelers have a better chance of seeing the McGee till. Start looking from Mammoth Creek on south.

● At Owens River-Airport Road, the white spot on the hillside at 3 o'clock is waste from an abandoned clay-processing plant that got its material from altered volcanic rocks in the hills to the north.

Inbound: The white spot is at 9 o'clock.

● Starting at the airport turnoff, knobs of dark lava appear close to the road on both sides. You are riding up onto the back of some partly buried lava flows coming out of the Mammoth embayment, into which you turn shortly. The ridges and hills right of the highway are composed of mostly older lavas that fill the Long Valley caldera.

Inbound: The extensive exposures of the lava on which Highway 395 runs are met just beyond the crossing of Mammoth Creek, nearly a mile from the cloverleaf. The hilly area to the left is underlain largely by lavas filling the 750,000-year-old Long Valley caldera.

● Within a mile the highway curves left, and you see clearly at 9:30 o'clock the bare, steep, "dump" moraine of Laurel Creek, traversed by a switch-back road. Good views continue up to and beyond our turnoff on Highway 203 to Mammoth.

Photo P–12.
Steam column and rows of condensers in mid-ground are part of Casa Diablo geothermal electrical generating plant just east of Highways 395/203 interchange. (Photo by Helen Z. Knudsen.)

● Approaching the Mammoth cloverleaf and exit (Highway 203), Casa Diablo Hot Springs lie 0.5 mile right at 1 o'clock. This is a geothermal area with natural steam vents where a good many test wells have been drilled to assay the steam resources for power generation, as now practiced at the "Geysers" north of San Francisco. During 1993, a considerable geothermal electrical development has been in operation here; three plants generate a cumulative 40 megawatts of electrical power. The plants draw hot water (330° to 350°F) from 13 shallow (500 feet) wells. This water vaporizes isobutane, which drives the turbines. The most visible structures are cooling condensers for the isobutane (Photo P–12). The spent geothermal water is pumped back into the ground by means of injection wells 2,000 to 2,500 feet deep.

Inbound: You should see steam plumes from the Casa Diablo operation approaching Highway 395 on 203 but nothing of the power plant unless you run under 395 to the end of the road.

● Upon leaving U.S. 395 you head west into the Mammoth embayment, and, within a few hundred yards, cliffs of black lava outcrop close by and continue on both sides of the highway for another mile. Two sequences of flows in these lavas have been dated to be approximately 200,000 and 300,000 years old. The older flows are separated from the upper younger group by an intercalated glacial deposit, the Casa Diablo till. Within this stretch along Highway 203 you see granitic boulders that have weathered out of the Casa Diablo till. The crest of the rise ahead is made by the younger sequence of lavas, which rest on the till and outcrop in road cuts approaching the top of the ascent.

Inbound: Granitic boulders from Casa Diablo till are seen for 0.6 mile, mostly left of the highway, after it drops over the crest of the descent and passes through the young lavas. The underlying older lavas extend to within 0.2 mile of the 395 on-ramp.

● A little over a mile from the 395 cloverleaf and just beyond the start of divided highway, you are riding over the younger lava flows. Within 0.5 mile large granitic boulders are resting on these lavas, and morainal ridges are crossed beyond Meridian Blvd. They were deposited during a relatively young glaciation.

The heavily timbered slope on the far side of the valley at 8-9 o'clock is the front of a massive bouldery moraine dumped into the Mammoth embayment by a glacier from Sherwin Creek. The Inyo National Forest visitor's center is on the right about a mile beyond Meridian Boulevard

Inbound: Between the visitor's center and Meridian Boulevard you pass through young glacial moraines with granitic boulders resting on top of the lavas seen descending the rise 0.4 mile beyond Meridian Boulevard. Views of Sherwin and Laurel creeks dump moraines to the south are excellent.

● For those driving on to Mammoth Mountain, the rocks north of the road beyond Mammoth Village are lavas composing volcanic domes. You will also see much young, fragmental pumice in road cuts, if they aren't covered by snow. This region has repeatedly been

blanketed by pumice falls thrown out of explosive vents in the Inyo and Mono craters to the north. Some of the pumice on Mammoth Mountain is 1,500 years old, as determined by a radio-carbon date on a buried tree stump. Explosion pits in the Inyo Craters, just a few miles north, formed only 650 years ago as shown by the same means, and some layers of explosive pumice there are as young as 250 years.

● The "earthquake fault" about two miles up the Mammoth Mountain-Devil's Postpile road, an open crevice in a lava flow, is not a good example of a fault. A fault is normally a tightly closed fracture along which crustal blocks have slipped past each other. Slippage is suggested here by some scratches on the walls of this crevice, but most of the movement seems to have been a splitting apart. Whether the splitting caused an earthquake is purely conjectural, because the event was prehistoric.

● Mammoth Mountain (see Photo P–9) is a complex volcanic dome. Some of its lavas are 400,000 years old. If you ski the north face of the hill at chair-lift 3 and go far enough east, you may spot a steam vent or two about half way down. The mountain is still warm and is best regarded as a dormant, not necessarily dead, volcano.

● Views from the high chairs and gondola at Mammoth Mountain are superb, especially west and north into the Middle Fork of the San Joaquin River, to the Minarets and the Ritter-Banner country. The low ridge extending north about 1.5 miles west of Mammoth Lodge is the drainage divide of the Sierra Nevada.

The Mammoth region is rich in geological phenomena and relationships. If you spend any time here and have an interest in hiking and natural history, get Genny S. Smith's excellent guide on the Mammoth Lakes. It's authentic, reliable, and well written. The visitor's center should have copies or know where you can get one.

1995: Horseshoe Campground has been closed because Trees are dying from Magma coming close (1mi) to the surface

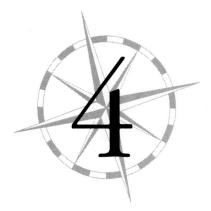

Supplementary Trip Guides

Six supplementary segments cover spurs off the basic Death Valley and Mammoth trips, as follows: Q, R, and S provide guides from the Los Angeles Basin to Palm Springs by way of San Bernardino and Riverside. Segment T deals with features along Interstate 5 (Golden State Freeway) from San Fernando Pass to southernmost San Joaquin Valley. Segment F does the same for the Baker to Las Vegas part of Interstate 15, and F_S describes a scenic loop into the Red Rock Recreational Area west of Las Vegas. All these segments except F_S are along freeways, and because freeway travel is speedy, you do well to review the appropriate segment texts in advance.

LAS VEGAS AREA

Segment F—Baker to Las Vegas, 89 Miles

● This segment starts at the Kelbaker Road overpass opposite the center of Baker (Figures 4–1 and 4–2).

▶ Record your odometer reading here.

Kelbaker Road extends 35 miles and more southeast to the old Kelso station on the Union Pacific Railroad. En route it passes alongside the extensive Cima volcanic field with its 65 lava flows and 50 vents. The road provides access to Kelso Valley, Kelso Dunes, and the Granite and Providence mountains in the heart of the East Mojave National Scenic Area.

Inbound: You see the Kelbaker Road country well descending to and passing Baker.

● From Baker you continue northeast on Interstate 15. The low bedrock hills a few miles directly south of town are the north end of Little Cowhole Mountain. They contain metamorphosed Paleozoic sedimentary rocks intruded by Mesozoic granitic bodies. The knobs, blackened

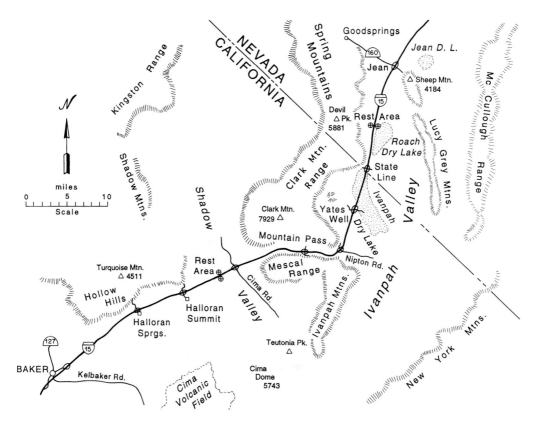

Figure 4–1.
Segment F, Baker to Jean (Nevada).

by rock varnish, have a patchy mantle of wind-blown sand and silt. At 2 o'clock on the right skyline are some of the many cinder cones of the Cima field (see Photo B–5).

Inbound: The Cowhole knobs are obvious at 9 o'clock passing Baker. The cinder cones are behind you; you need to stop to see them.

Beyond the Main Street overpass, the highway straightens for the 19 mile, 3,000 foot climb to Halloran Summit. A curve to the right short of the Summit puts it out of sight. The ascent looks deceptively gentle, but it steepens toward the head, and drivers should heed warning signs to turn off air conditioners on hot days. Many cars boil or vapor lock on this grade. A special truck lane starts 0.5 mile ahead.

Inbound: No problem with air conditioners coming down.

● At "*Las Vegas 89, Salt Lake City 534,*" high peaks, including Turquoise Mountain (4,398) in the rough, ragged Hollow Hills are at 11 o'clock. Bedrock there consists of Precambrian metamorphics intruded by Mesozoic granitics. The ragged terrain reflects the heterogeneity of the Precambrian rocks and contrasts sharply with the smooth, even skyline at 12:45–2:30 o'clock created by young Cima lava flows. Hollow Hills are pockmarked by prospects and abandoned mine workings, but no major productive mines were developed.

Inbound: At 16 miles from Halloran Summit you see these Hollow Hills features at 1–2 o'clock.

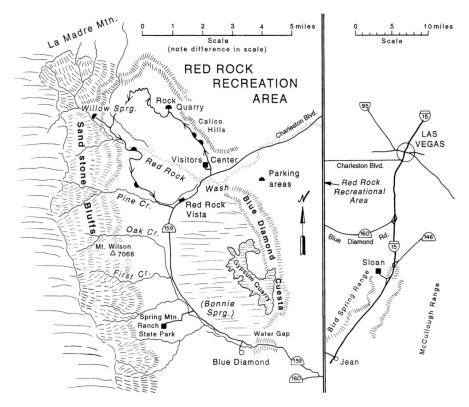

Figure 4–2.
Jean to Las Vegas and Red Rock Recreational Area (Segment F_S).

● Your highway is ascending a smooth, largely undissected, gently sloping (2° to 3°) alluvial surface of disintegrated granitic debris with essentially no cobbles or boulders. Highway construction was easy and relatively inexpensive. No major cuts or rock excavations were required. Berm-like ridges have been bulldozed locally to control floods.

At 7 miles from Baker the prevailing creosote-bush vegetation is supplemented by salt cedar and palos verdes trees along the highway. These are not native to this area and have probably re-seeded from plantings made at a former rest area site 0.1 mile beyond *"Elevation 2,000 Ft."* Joshua trees, which are native, appear shortly and increase in abundance to Halloran Summit, beyond which they make a veritable forest.

Inbound: The site of the former rest area is south of the highway, 8.3 miles from Halloran Summit.

● Around 1.7 miles beyond Halloran Springs overpass and *"Elevation 3,000 Ft."* at 10 o'clock is a 40 foot cliff of Cima lava tilted gently west resting on a smooth surface that truncates underlying pale pinkish fanglomerate beds tilted about 10° east (Photo F–1). This lovely angular unconformity shows that the fanglomerates were deposited, tilted, and eroded to a smooth, nearly flat surface before the 5 m.y. old lava was extruded. The fanglomerates are Miocene and the lavas earliest Pliocene. The unconformity thus marks the break between the Pliocene and Miocene epochs of the Tertiary period (see Appendix A). Following extrusion, the lavas, fanglomerates,

Photo F–1.
Angular unconformity between Pliocene (5 m.y.) lava and tilted Miocene fanglomerates in north wall of Halloran Wash less than two miles north of Halloran Springs overpass . (Photo by Helen Z. Knudsen.)

and unconformity were all tilted west about 10°. This exposure reveals a nice bit of geological history for you to read.

Inbound: These relationships are seen at 2–3 o'clock, 4.5 to 5 miles from Halloran Summit.

● Looking ahead, the lava cliff extends about 2.5 miles to a dark lava knob that looks like an intrusive plug, but is probably just an unusually thick accumulation of surface lavas. This knob is at 3 o'clock near *"Elevation 4,000 Ft."*

Inbound: The lava knob is 2.5 miles below Halloran Summit.

● Once you have driven over Halloran Summit, an extensive vista opens, dominated by Clark Mountain (7,928) on the 11:30 o'clock skyline (Photo F–2). The highway extends straight ahead, descending into Shadow Valley and ascending its far side toward Mountain Pass (4,730), which separates the Clark Mountain Range from the Mescal Range and Ivanpah Mountains that extend far south. Most of the mountains in view consist principally of Paleozoic carbonate beds, *limestone* and *dolomite*. Joshua trees are abundant beyond Halloran Summit.

Inbound: About the only way inbound passengers can sample this vista is to stop on the Halloran Summit off ramp. It shows you where you have been.

● Descending from Halloran Summit, the southern skyline at 2 o'clock features a broad, smooth, rounded dome of granitic rock, Cima

Photo F–2.
Northeastward view from Interstate 15 of Clark Mountain, across Shadow Valley. (Photo by Helen Z. Knudsen.)

Dome (Photo F–3). *Cima* means "summit" in Spanish. The dome is deceptive, because it is fully 700 feet higher than Halloran Summit. Because of its unusual form, it has received considerable attention from geologists for many decades. It is probably a tectonically upwarped part of an old, smooth, nearly flat erosion surface, formerly mantled at least in part by Cima lavas. The smooth plane truncating the tilted fanglomerates at the Halloran grade unconformity (see Photo F–2) is probably a part of this same surface.

The dome is clearly seen descending into Shadow Valley, and views are especially good in the vicinity of the Cima Road exit (see Photo F–3). Beyond Cima Road, Teutonia Peak is clearly evident on the dome's northeast flank. The peak is an erosion residual of unusually resistant granitic rock not worn down to the level of the old surface.

Inbound: Cima Dome and Teutonia Peak are in good view to the left as you approach Cima Road.

● As you near the rest area 5.3 miles from Halloran Summit, Clark Mountain looms ever higher and about 1.5 miles ahead are exposures of white deposits on both sides of the road. In passing Cima Road you see these shallowly dissected, white, fine-grained beds close up. They are said to be of lacustrine and fluvial origin, which accumulated partly in a small pluvial lake occupying the floor of Shadow Valley about 10,000 to 20,000 years ago. The deposits contain scattered fossil bones of elephants, camels, small horses, and rabbits living here at the same time animals were getting trapped in the famous La Brea tar pits of Los Angeles.

Inbound: The white lake beds are seen approaching and passing Cima Road junction.

● On the upper part of the ascent into Mountain Pass, rocks on either side are mostly Paleozoic carbonate layers. Near and beyond *"Elevation 4,000 Ft."* the scars and waste piles of abandoned prospects and mines on Mohawk Hill to the left, and farther ahead on the right in

Photo F–3.
Cima Dome looking south from Cima Road overpass of I-15. Teutonia Peak on dome's east flank.
(Photo by Helen Z. Knudsen.)

the Mescal Range, are numerous. None of these mines were very large. Gold and silver were the principal objectives, but the ores also contained some lead, zinc, and copper.

Inbound: Beyond "Baker 33, Barstow 95," you see prospects and mines mostly right of the highway. Where the route straightens for the run into Shadow Valley, a nice vista lies ahead.

● At *"Brake Check Area 2 Miles"* unusually good stratification exists in limestone beds left of the highway. Just beyond *"Brake Check Area 1/2 Mile,"* the first of several large waste piles of a current mining operation comes into view at about 11 o'clock. Other waste piles and some buildings are seen as you progress toward the summit of Mountain Pass (4,730). A short detour off onto Bailey Road gives a more comprehensive view of this operation (Photo F–4). At the stop sign turn right to park and make a short walk to a vantage point.

Inbound: You pass the mining operation too quickly to see much, but a short detour and pause at the top of the Bailey Road off ramp can remedy that.

● This is truly an unusual mine. It provides over 25 percent of the world's rare-earth elements and harbors one of the world's largest proven reserves. More than seven valuable rare-earth elements are recovered here, with cerium and lanthanum the most abundant. Rare earths are employed in a variety of processes and products such as alloys, high-temperature superconductors, special glasses and ceramics, absorbing filters, petroleum cracking, microwave control devices, x-ray screens, magnets, camera lens, lighter flints, exhaust catalysts, flourescent lamps, tape drives, and other special uses. Of interest to all of us is their part in making color television possible.

● The ore mineral is a complex rare-earth fluorcarbonate, *bastnaesite,* that occurs in an unusual igneous rock, *carbonatite,* which has been intruded as a *sill* into Precambrian metamorphic rocks. The ore body is about 2,400 feet long, 200 feet thick, and extends at least 500 feet deep. The ore is mined in open pits, crushed and treated on the site.

Photo F–4.
Looking north to buildings of Mountain Pass rare earths mining operation from Bailey Road overpass of I-5. (Photo by Helen Z. Knudsen.)

The discovery was made in 1949 by two seasoned prospectors searching for uranium and thorium with a geiger counter. They collected a rock with an unusually high reading and submitted a sample to an expert geologist. He recognized the bastnaesite, had it analyzed, and proved it to be exceptionally rich in rare earths. He advised the prospectors to forget uranium and thorium and to stake their claims firmly on the carbonatite rock, which contained the bastnaesite. They did and later sold the claims for a modest sum, considering their subsequent proven value. Ownership of the property in 1993 resided with Molycorp and Unocal (Union Oil Company of California). The mining is done in open pits. Huge quantities of crushed rock are processed locally, and further refining of the concentrate is done on the site.

▶ | Note your odometer reading leaving Bailey Road.

● Within 1.5 miles, just beyond *"Runaway Truck Ramp 3 Miles,"* the highway enters a deep, double-walled road cut through old, intensely squeezed and sheared Precambrian metamorphic rocks, mostly gneiss. These are part of the complex into which the rare-earth-bearing carbonatite was intruded. Views of these rocks are best in the left wall of the cut. Beyond, the highway descends Wheaton Wash, a narrow steep walled canyon cut in hard Precambrian rocks all the way to the foot of the grade.

From upper reaches of Wheaton Wash you can get sweeping views out across long, wide Ivanpah Valley to the Lucy Gray Mountains on its east side. The rough, jagged skyline of the Lucy Grays, at about 2:45, reflects the heterogeneity and complex structure of the Precambrian metamorphic rocks composing those mountains.

Inbound: At "Baker 43, Barstow 109, Los Angeles 222," the ragged crest of the Lucy Gray Mountains is on the skyline at 9 o'clock. You enter Precambrian rocks in lower Wheaton Wash about 0.7 mile beyond the Nipton overpass. In 2.5 miles road cuts give good exposures of Precambrian metamorphics. The big, double-walled road cut is just beyond "Bailey Road Exit, 1 Mile."

● At the Nipton overpass the highway straightens into a north-northeasterly course up the floor of Ivanpah Valley. Ivanpah Dry Lake lies ahead, and you cross some of it one mile beyond the Yates Well overpass, which is five miles from Nipton Road. Between Nipton and Yates overpasses Clark Mountain Range lies to the left. Its southern part consists of the Precambrian complex, but the northern part is made up of well-layered, deformed Paleozoic sedimentary rocks. The contact between these two units is visible at 9 o'clock approaching Yates Well overpass. See whether you can spot it. The large dark knobs well out on the valley floor at 9 o'clock beyond the overpass are erosion residuals held up by especially resistant rocks within the Precambrian complex.

Inbound: Yates Well overpass is a good point of reference in looking for these features.

● The steaming plant a mile or so left of the Yates Well overpass is a coal-burning electricity-generating operation, which uses rejected coal trucked in from the large Mojave generating plant at Laughlin, Nevada, on the Colorado River. Why is it located here? Probably because there is lots of room, land is cheap, ground water is adequate, no major habitation is nearby, and the product goes easily into the local electrical power grid.

● After crossing a part of the Ivanpah Dry Lake, a playa, you come to gambling establishments at State Line, Whiskey Pete's on the left and Prima Donna on the right, designed to divert weary Californians from continuing to Las Vegas. A similar arrangement exists at Jean, 12.3 miles ahead. The rock spur just behind and beyond Whiskey Pete's consists of cavernously weathered, rather messy lowermost Paleozoic limestone and dolomite beds.

Inbound: These features don't need any navigational aids.

> ▶ Record your odometer reading at State Line.

● Beyond State Line the southern part of the Spring Mountains lies left. Like the northern Clark Mountains with which they are contiguous, the Spring Mountains consist mostly of deformed, well-stratified Paleozoic formations. The far high peak at 11 o'clock with a road up its south flank is Potosi Mountain (8,512). To the right is the northern part of the Lucy Grays and Roach Dry Lake, which nearly connects with Ivanpah Dry Lake. Both dry lakes were undoubtedly submerged by pluvial Ivanpah Lake. At that time it would have been possible to make the 27 mile journey from Nipton to Jean by boat. That would have been fun. The shores would have been lined with tules and populated by ducks, geese, and a variety of other shore birds. Native Americans might have waved to you from the beaches.

Inbound: These features and relationships are seen after leaving Jean and before State Line, which is about where the highway disappears ahead.

● About 4.5 miles from State Line is a rest area, a good site for a leisurely inspection of the surrounding terrain. To the east across Roach Dry Lake are the Lucy Gray Mountains, consisting of Precambrian rocks.

To the west at 9 o'clock within the Spring Mountains is sharp, pyramidal Devil Peak (5,881) looking different than the Paleozoic rocks on either side. Its rocks are more massive, more homogeneous, and unstratified. It is a fine-grained, near-surface igneous intrusive body of *rhyolite,* 25 m.y. old. Related intrusions are thought to be responsible for the mineralization around the old mining camp of Goodsprings (Mr. Good's springs) northwest of Jean.

Inbound: Views as you approach and reach the rest area are the same and as good as outbound. Devil Peak is at 3 o'clock, and Roach Dry Lake is more obvious to the east. Potosi Peak lies behind at 4:30 o'clock and the small roadside knobs of Paleozoic limestone are just beyond "State Line 11, Baker 76," about 1.5 miles south of Jean.

● Continuing north, at 4.5 miles from the rest area, little bedrock knobs just left of the highway give a close look at Paleozoic limestone beds. At Jean the Goldstrike and Nevada Landing (sidewheelers) resort-gambling complexes border the highway (see Figure 4–2). The assemblage of buildings, fences, and light towers near the base of Sheep Mountain behind Jean is a Nevada state prison. Rocks in the mountains are Paleozoic strata tilted gently eastward.

Highway 161 to the old mining camp of Goodsprings (6.2 miles) goes northwest 0.6 mile beyond the center of Jean. Goodsprings was principally a silver-lead operation, and in its heyday supported a fair-sized population. The town has not been gussied up for tourists and is worth the short detour. The 50 people currently living there appreciate its antiquity.

Inbound: If you are visiting Goodsprings, watch for the 161 turnoff before the center of Jean.

● Three to 4 miles beyond Jean, low hills left of the highway are part of the Bird Spring Range. The extensive exposures of thinly bedded limestones are characteristic of the easily recognized marine Permian Bird Spring Formation, about 250 m.y. The beds are folded as shown by the different degrees and directions of inclination *(dip).*

Inbound: Bird Spring Hills are on the right at the end of the concrete pavement.

● A section of concrete pavement roughly 10 miles long starts about 4 miles beyond Jean.

There, foothills of the McCullough Range to the right have a considerable mantle of light-colored, fine, wind-blown sand and silt probably derived from Jean Dry Lake, located east and a little north of that settlement. These deposits constitute climbing dunes. Wind is one of the few natural agents of transport that carries material uphill.

Inbound: These climbing dunes are seen to the left at 8 o'clock beyond the end of the concrete highway.

● In the next 3 to 5 miles outcroppings of Paleozoic rocks, mostly limestone, are close alongside the highway to the left. In places road cuts through rock knobs give good exposures of the beds. The highway crosses the Union Pacific Railroad on an overpass, 9 miles beyond Jean. Starting about 1 mile beyond the overpass, outcrops of dark volcanics lie left of the highway and extend for nearly a mile. At *"Check Station, All Trucks, 1 Mile,"* look to 10:45 o'clock to see quarries in massive dolomitized (magnesium added) Mississippian (350 m.y.) limestone that supply the cement plant at Sloan. The Union Pacific tracks curve that way to maintain a level route, much to the pleasure of the cement company (U.S. Lime Products). Beyond the Sloan exit near *"Las Vegas 13, N. Las Vegas 15,"* the concrete highway ends and your route starts a gradual curve left to pass through a natural saddle in a projecting ridge of massive limestone, the same rock quarried at Sloan. Beyond the saddle, the route is headed due north toward Las Vegas.

Inbound: Southbound you pass through the saddle 0.7 mile south of the Henderson-Lake Mead interchange, and in 1.6 miles you are at the Sloan exit (No. 25). Look back to 4 o'clock to see the Sloan quarries. The volcanics start on the right about 2.5 miles beyond the Sloan exit. The little roadside knobs of limestone start less than a mile beyond the overpass across the railroad.

● Beyond the Henderson/Lake Mead interchange, you traverse the floor of Las Vegas Valley. The site of Las Vegas was originally a meadow on the old Spanish Trail to California, much used and appreciated by pioneer parties recuperating before tackling the desert country ahead.

As you proceed toward Las Vegas, some towering sandstone cliffs making the western skyline are impressive. Locally known as the Sandstone Bluffs, they are composed of the Jurassic (200 m.y.) Aztec (Navajo) sandstone, the same formation that makes the walls of Zion Canyon, Utah. If lighting is favorable, you may be able to see that the bluffs are capped by a thin layer of darker rock, a dolomitic-limestone formation about 500 m.y. old. If you are puzzled that old rocks rest on top of younger rocks, read part of the supplementary F_S segment that follows. Better still, make the trip it describes.

At 1:30 o'clock behind Las Vegas is free-standing Frenchman Mountain, beloved by geologists because it contains a complete and continuous section of Paleozoic and Mesozoic rocks of the regions. It is like a reference library.

Inbound: The Sandstone Bluffs are at 2–4 o'clock, but Frenchman Mountain, at about 7:30, is difficult to see unless you stop for a look back.

● This segment terminates at the Blue Diamond/Pahrump interchange, about 5 to 8 miles from central Las Vegas. Those going ahead will have other things on their minds. Those intrigued by the great Sandstone Bluffs can turn off on Highway 160 here and do a loop trip through the Red Rock Recreational Area ending up in Las Vegas after a 30 mile swing. Segment F_S describes this journey.

Segment F_S—Red Rock Recreation Area, (Nevada), 34 Miles

Travelers willing to take a roundabout (see Figure 4–2) but highly scenic loop to Las Vegas should turn off Interstate 15 at the Blue

Diamond/Pahrump off ramp and at the first boulevard stop go left on Highway 160 (Blue Diamond Road).

> ► | Record your odometer reading here.

● The first few miles are across alluvium burying Paleozoic rocks of the type seen since Clark Mountain. You are headed directly toward the 2,000 to 3,000 foot, massive, Sandstone Bluffs. With a morning sun and clear weather, you should be able to see that the white and red sandstone cliffs are capped by a darker rock. It is a sheet of dark gray limestone and dolomite of late Cambrian age (525 m.y.) shoved eastward over the top of the younger Aztec sandstone (200 m.y.) along the low-angle Wilson thrust fault. Views of this relationship are best at about 5 miles west on Blue Diamond Road. Farther west you get too close under the cliffs to see this relationship, except in one or two places. In a normal sedimentary sequence, younger rocks rest on top of older rocks; the relationship here is one of the commonest ways of reversing that order. The Wilson is just one of several large thrusts in this region.

Inbound: To get a satisfactory view of Wilson fault relationships, you will probably have to stop, get out, and look back about 5 miles east from the 159/160 highways junction.

● As you continue west, you become aware of a high, forelying, linear ridge with a steep cliffy face to the east extending right (north) from your route. This is Blue Diamond Hill (see Figure 4–2), and eventually you learn that it is strongly asymmetric; has a steep cliff faced to the east; smooth, and a much gentler backslope to the west (which is conformable with bedding in the rocks). Ridges like that are called *cuestas* (Figure 4–3), a mellifluous Spanish word. Were the ridge about equally steep on both sides, it would be called a *hogback* (see Figure 4–3).

Inbound: You have already seen the back, so-called dip-slope, side of Blue Diamond Hill,

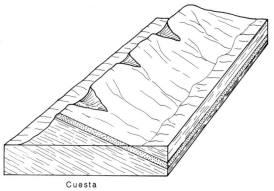

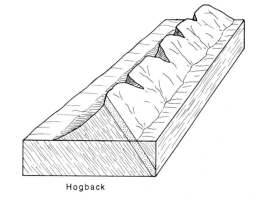

Cuesta   Hogback

Figure 4–3.
Sketch of a cuesta and a hogback.

and can see the steep front cliff face looking back at 7–8 o'clock from Highway 160 going east from its junction with 159.

● At 10.5 miles from I-5, turn right onto Highway 159 and proceed northwesterly toward a gap in the Blue Diamond Hill cuesta.

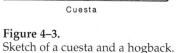

▶ | Record your odometer reading here.

That opening was cut by the stream which flows through it, so it is called a *water gap*. If the gap had been abandoned by the stream and came to lie well above stream level, because of erosion or tectonic uplift, it would be known as a *wind gap*. Pennsylvania has many water gaps, where large rivers such as the Delaware or the Susquehana cut across ridges of the Valley and Ridge province.

Approaching the 160/159 intersection you may have observed white areas at the top of Blue Diamond Hill. They are quarries (see Figure 4–2) in Permian sedimentary rocks from which gypsum ($CaSO_4 \cdot 2H_2O$) is recovered for use in the plant ahead. Gypsum is used to make plaster and such products as wall board. The plant and quarries were operated for many years by Blue Diamond Company and only recently taken over by James Hardy Gypsum.

The gypsum indicates that a dry climate ruled here for at least part of the Permian Period. Just 1.2 miles beyond the 160/159 junction, the gypsum plant is directly ahead.

Inbound: White quarry scars at the crest of Blue Diamond Hill may catch your eye at several places as you pass around the south end of Blue Diamond Hill and go through the water gap. The gypsum plant is just beyond on the left.

● About 0.7 mile beyond the gypsum plant, you enter the water gap. The steep walls are made by massive Kaibab and Toroweap limestone layers. These are the same layers that make the top of the east face of Blue Diamond cuesta and the rim of the Grand Canyon at El Tovar.

Upon emerging from the water gap, you get a splendid view of the Sandstone Bluffs. Within a mile near "*Bonnie Spring 2, BLM Visitor's Center 8,*" you see that the base of the sandstone cliff is underlain by 100 to 200 feet of deep red, soft shale and mudstone of the Triassic (250 m.y.) Moenave Formation. This is an important relationship. Massive sandstone can form a cliff, but if the cliff is to remain fresh and nearly vertical, it must be continually undermined (sapped) at the base. The soft red beds perform that function. You will see more of them farther along.

Inbound: Red beds at the base of the Aztec cliffs are seen at several places to the right before you enter the water gap.

● At 3.8 miles from the 160/159 junction, you enter the Red Rock Recreation Area. Within 0.3 mile farther, on the right you see good exposures of gently dipping, well-layered fawn colored limestone beds, the Virgin Spring member of the Triassic Moenkopi Formation. It is the basal unit of the Mesozoic sequence and forms the lowermost part of the Blue Diamond cuesta's back slope.

The early Mesozoic landscape was a low, featureless plain in this area with a shallow sea on its edge. Slight changes in either sea level or land elevation caused large shifts in shoreline position. Consequently, marine and land-laid beds commonly interfinger here within the Moenkopi Formation. The Virgin limestone represents a time when the land was under the sea. Careful searching may reveal marine fossil shells in these beds.

Inbound: In 0.8 mile beyond the Bonnie Spring turnoff you come to good exposures of the Virgin Spring limestone on the hillsides to the left. This is the bottom of the back-slope of Blue Diamond Hill cuesta, to be seen in its entirety ahead.

● Shortly the road to the "Old Nevada" resort, site of mock-western gun fights, turns off left, and in another 0.7 miles is the entrance to Spring Mountain Ranch State Park (Bonnie Spring), a delightful place to picnic. Brilliantly yellow cottonwoods grace this area in the fall. Joshua trees are starting to appear, and juniper trees add to the scene a little farther along.

Beyond Spring Mountain Ranch junction, large angular blocks of sandstone to the left along the base of the bluffs should attract your attention. Some fell as individual pieces from the cliff, some are remnants of ancient rock fall-slides, and some have been carried well out from the cliff by debris flows that have a high capacity for transport. *Debris flows* are masses of broken-up rock detritus, usually in an abundant muddy ma-

trix, that flow downslope on hillsides or in canyons. You can see that the bluffs are indented by deep steep walled canyons from which intermittent streams carry and spread sand and gravel over the soft Triassic beds underlying the low area between the base of the bluffs and the back of Blue Diamond Hill cuesta.

A little more than a mile from Spring Mountain Ranch is the road into Oak Creek Canyon and the only campground in the area (in 1993). At this time it was not a well-organized or equipped camping area. Near the bluffs, the Oak Creek road is rough with high centers not suitable for modern touring cars. About 0.2 mile beyond the Oak Creek road you see at about 8:45 o'clock south of Oak Creek, a curious large forelying sandstone knoll standing in front of the base of the bluffs. It is an old, large landslide block. (Photo F$_S$–1)

The walls of Oak Creek Canyon where it cuts back into Sandstone Bluffs are locally as high and as grand as the walls of Zion Canyon, although they lack the arches and other sculptures. Driving north on Highway 159, alert observers will catch occasional glimpses of the dark Cambrian beds above the Wilson thrust capping the Sandstone Bluffs.

Inbound: Enjoy the views of the Sandstone Bluffs driving south. The various side roads turning off Highway 159, all to the right, are well labeled.

● As the highway curves northeasterly, the dark red Aztec sandstone of Calico Hills and Turtlehead Mountain lie ahead at 10-11 o'clock. Opinions differ as to whether the Aztec was originally all red, all white, or mixed. If you choose to take the one-way loop drive of 13 miles from near the Visitor's Center, you will have opportunity to observe relationships bearing on this matter. A plaque at the Red Rock overlook ahead says that the Aztec was originally all red, and the white parts have been leached. This is probably the right interpretation.

Photo F$_S$–1.
Sandstone bluffs in Red Rock Recreational Area at Mt. Wilson, local relief a good 2500 feet. (Photo by
Helen Z. Knudsen.)

Within 1.5 miles from Oak Creek Road you
pass the outlet of the one-way loop road, and
0.5 mile beyond is the left turn into Red Rock
overlook, a stop worth making for the view
alone. A diorama identifies features of interest
including Mt. Wilson (7,068), not the highest but
one of the most imposing peaks of the area. It
gave its name to the Wilson thrust. With proper
lighting and visibility this is a fine photo spot.

*Inbound: Localities are so well marked or obvi-
ous along this reach of Highway 159 you don't
need navigation pointers.*

● In 1.7 miles beyond the outlook is the turn-in
left to the BLM Visitor's Center, a worthwhile
stop. It has maps, publications, and useful information on the Red Rock area, plus excellent
restrooms. Be aware that the Center's parking

area closes at 4 p.m. (unless changed). The 13
mile, one-way loop, drive starts from nearby
and can be traveled from 8 a.m. to 5 p.m. daily.
Get a leaflet at the Visitor's Center showing
loop road's configuration, points of interest,
and parking spots. It is a good drive and gets
you close to Aztec sandstone exposures.

Travelers wishing to proceed directly to Las
Vegas do so best by reentering Highway 159
and proceeding east (see Figure 4–2) 12.5 miles
via Charleston Boulevard to I-5 near the Strip.
Charleston Boulevard has many stoplights, so
take a patience pill if you have one.

*Inbound: Inbound travelers can take the one-
way loop drive and recover Highway 159 just
south of the Red Rock overlook, or they can
take 159 southwest from the Visitor's Center.*

Photo F$_S$–2.
Cross bedding in Aztec sandstone of Calico Hills. (Photo by Helen Z. Knudsen.)

> ▶ Record your odometer reading at the gate of the loop drive entrance.

Anyone traveling the one-way loop in reverse is in trouble, and no inbound navigational aids are furnished.

Also be aware of and keep track of the roadside mileage posts (MP). These are truly posts, not markers (MM). They are placed every mile along the route and are useful locators.

● Just 0.8 mile out on the loop road is a good parking area on the right at the base of the Aztec cliffs of the Calico Hills, so named for their irregular mixture of red and white. Here you see excellent examples of cross bedding (Photo F$_S$–2); that is, bedding inclined in various directions from the expected normal stratification. To many (but not all) geologists, cross bedding like this indicates that the Aztec accumulated as a series of superimposed sand dunes. If you look closely you will see places where the cross beds are convoluted into small, complex folds (Photo F$_S$–3). This is interpreted to be the result of slumping. Dry sand slumps frequently on the lee side of transverse dunes, but it doesn't make convoluted structures. The sand has to be coherent, and the best way to do that is to have it deeply wetted by rain or deposited in an aqueous environment. Some geologists feel that convolutions of the type seen here and at the next stop show that the Aztec (Navajo) sandstone accumulated on the sea floor, not on the land as dunes. This disagreement is yet to be resolved. This geologist thinks it possible that the cross beds are of eolian origin and that the slumping to produce convolutions in them occurred following un-

Photo F$_S$–3.
Convoluted (slumped) cross bedding in Aztec sandstone, Calico Hills. (Photo by Helen Z. Knudsen.)

usually heavy rain, which wetted the sand deeply, gave it coherence, and added weight that caused slumping.

● In another 0.6 mile is a large parking area at Calico Hills overlook. The display of cross bedding and convolutions here is possibly even better than at the first stop. Most of the stops shown on the map (see Figure 4–2) and the BLM leaflet are identified by signs and have features of interest. This discussion doesn't comment on all of them but calls attention to other points of interest in between.

● About 0.3 mile beyond Calico Hills overlook, at the north end of Calico Hills, park alongside the road and study relationships between white and red sandstone. Note the remarkably sharp contact between areas of different color, but also note that such contacts cut across the bedding, which shows that the present color arrangement postdates deposition of the sandstone. By looking carefully at red areas you will see they harbor a number of sharply defined, small white spots. A comparable area of white sandstone shows no such red spots. All these relationships could be established by local leaching of a sandstone that was originally of uniform color. It is a lot easier to make small white spots in red sandstone by leaching than to convert a large area of white sandstone to red by staining through deposition of iron oxide, leaving little white spots of the type seen here. The weight of the evidence favors an original all red Aztec (Navajo) sandstone.

● The sandstone quarry turn-in 0.3 miles farther has a large parking area with rest rooms and a trail leading about 100 yards to the quarry site. This operation ceased in 1912, and the size of the blocks and the labor involved is impressive. Many people enjoy climbing among the sandstone knobs beyond the quarry.

In another two miles is MP 5 beyond which the highway corkscrews down into deep gullies and back out onto alluvium mantled spurs. At this point you see to the right dark Cambrian beds that lie above the Wilson thrust. Approaching and beyond MP 5 you must have noted how white the alluvial gravels are in road cuts. This is because of secondary deposits of calcium carbonate ($CaCO_3$) in the gravels that is developed by weathering. This sort of deposit goes by the Spanish term *caliche*. The calcium carbonate comes largely from limestone fragments in the gravels and lime-rich, windblown dust. In places it makes platy layers of solid $CaCO_3$, tough enough to ring under a hammer. Locally, the gravels are firmly cemented by the caliche deposit.

The White Spring parking area turnoff comes just beyond MP 6. From here on you see many exposures of soft red shale and mudstone beds at the base of the Aztec sandstone that undermine its cliffs, keeping them fresh and steep. In 0.7 mile at the bottom of a major gully, and the tip of a long switchback to the right, you get a closeup look at the red beds.

In just another 0.3 mile, where you top out, coming up out of the big gully, is an unusual double-walled road cut. With care you can park just beyond on the right and walk back for a closer look. In the outside cut you see a jumble of large, angular sandstone blocks stained an orange-yellow by weathering. Near the top they are a rich-brown color owing to more intense weathering closer to the surface. On top of the jumbled blocks is a relatively thin layer of gray stream gravel containing many rounded lime-

stone cobbles. Now turn and look at the inside road cut. There the gray gravel is thicker and relationships clearer (Photo F_S–4). What you have here is a small remnant of an old rock-fall slide, the jumbled angular blocks from the Sandstone Bluffs that occurred before the gully you have just crossed was cut. The gray stream gravels were deposited by the intermittent stream of the gully before it became entrenched. The stream at that time flowed just above the level on which you are standing. The orange-yellow staining of the blocks and their dark brown color at the top indicate a considerable episode of weathering before deposition of the stream gravel. Geologically, the slide was a young event, but in our chronology it had considerable antiquity, perhaps 20,000 to 30,000 years or more.

● If you drive into the nice Willow Spring picnic area for lunch, watch the left wall of the canyon. It once had an extensive coating of dark rock varnish, much of which has now weathered away giving the cliffs a scabby appearance. The varnish is destroyed because the rock underneath becomes weathered and flakes off, not because the varnish itself deteriorates.

● One tenth mile beyond MP 8 is a parking area on the right serving a trail head and a good overlook. The foreground of the cliffs here displays a number of large sandstone blocks delivered from the cliffs by free fall, rock-fall slides, or debris flows. Two miles farther along, 150 to 300 feet of the basal red beds are in good view. In another 0.7 mile is the turn-in to the parking at Pine Creek overlook, a trail-head site with rest rooms and an impressive view south to Mt. Wilson. Look at the huge boulder of cemented breccia in the parking area for an appreciation of effective caliche cementation. From the Pine Creek parking to junction with Highway 159 you travel across alluvium.

Photo F$_S$–4.
Rockfall-slide debris, from Aztec sandstone, overlain by younger, carbonate-stone rich, fluvial sand and gravel, on Red Rock loop road, about one mile beyond White Spring turn off. (Photo by Helen Z. Knudsen.)

● At the loop road/159 intersection, Las Vegas travelers turn left for their 16 mile trip. Inbound travelers may wish to go left, too, for 0.5 mile to the Red Rock overlook left off Highway 159, before starting south and east on their 20 mile trek to Interstate 5 at the Blue Diamond/Pahrump interchange. They need to shuffle back in the text of outbound Segment F$_S$ to the part dealing with the stop at the Red Rock overlook and read back from there.

Many Californians coming to Las Vegas have no idea that a place like the Red Rock Recreational Area exists. We hope you found it unexpected, scenic, and interesting.

Inbound: Travelers headed home from Las Vegas but intrigued by the Red Rock Recreational Area should get onto Charleston Boulevard and proceed west to the Red Rock BLM Visitor's Center. There they can pick up the pertinent part of Segment F$_S$.

LOS ANGELES BASIN TO PALM SPRINGS

A Y-shaped arrangement of Segments Q, R, and S (Figure 4–4) covers the most heavily traveled routes to Palm Springs.

Segment Q—San Bernardino to Beaumont, 21 Miles

● This segment extends eastward on I-10 from the interchange with I-215 near the southwest corner of San Bernardino (see Figure 4–4). Travelers approaching this point from the west on I-10 should find the following features of interest. East of the I-15 interchange, the route crosses a sandy plain composed of weathered and disintegrated rock debris swept down by streams flowing from the San Gabriel Mountains 8 miles north. Sand blown by strong Santa Ana winds can make this stretch unpleasant to impassable. Old stabilized dunes are cut through by the freeway on the left going into Colton east of Riverside Avenue. Sunset Dunes Golf Course, on Valley Boulevard, is just to the north. The steep south face of the San Gabriels, rising nearly 7,000 feet from its foot at Ontario (8,693) and Cucamonga peaks (8,856), was formed by repeated displacements along the Cucamonga fault zone at the range base.

Inbound: Just after the highway straightens into a due westward course coming out of Colton three successive road cuts on the right are in old dunes.

● At and beyond *"Rialto City Limit,"* eastbound travelers get a view on a clear day of high San Bernardino Peak (10,649) at 11:45 o'clock. San Bernardino Peak is the western end of the eastern San Bernardino Mountains that lie somewhat south and east of the lower western part of the range, seen at 9:30–11 o'clock. San Bernardino Peak is an important reference point in southern California for the following reason.

After acquiring its vast western territories, the United States knew that a systematic land-survey system was needed. The government chose to divide the land into squares of two sizes; the larger, six miles on a side, are known as *townships,* and the smaller, one mile on a side, became known as *sections.* Each township contains 36 sections, and each section consists of one square mile or 640 acres. Location of townships within a reference grid was accomplished by establishing north-south and east-west reference lines, usually one or two within each state. The east-west lines are called *base lines,* and one of the two base lines for California passes almost through the top of San Bernardino Peak. Well-known Base Line Road is an east-west street four miles north of I-10. It runs along the San Bernardino Base Line, which continues westward about under the Huntington Hotel in Pasadena and out to the coast near Port Hueneme.

Inbound: Unfortunately, San Bernardino Peak is behind you here, but it was seen many times farther east, between Beaumont and Calimesa.

● As you travel east along I-10, between Rialto and Colton, you can glimpse through the trees the abrupt southern face of the western San Bernardino Mountains to the left at 9:15–9:30 o'clock toward the northeast skyline. This linear face is determined by the San Andreas fault, which lies at the base of the mountains after having slashed east-southeastward between the San Gabriel and San Bernardino Mountains along Cajon Creek and Interstate 15 (see Segment A). Rocks composing the San Bernardinos are largely relatively old igneous and metamorphic crystalline varieties, similar to, but not identical with, rocks in the San Gabriels.

Inbound: You have seen this southwest face of the San Bernardinos better at 2–3 o'clock between Redlands and I-215.

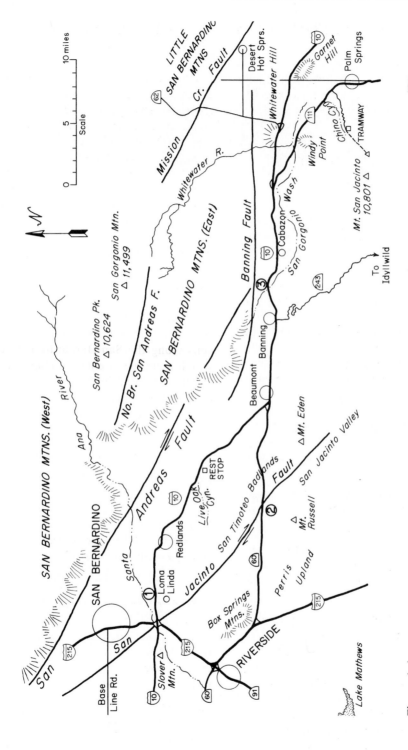

Figure 4-4.
Segments Q, R, and S, San Bernardino-Riverside to Palm Springs.

Photo Q–1.
Slover Mountain from the west, much modified by cement-company quarries in its marble. Perhaps the only mountain in California possibly destined to become a hole. (Photo by Helen Z. Knudsen.)

● At and to the east of Pepper Avenue, Jurupa Mountains are in view a mile or two ahead on the right. They are part of the northwest-trending Peninsular Ranges, a province (see Figure 2–1) distinct from the east-west trending San Gabriels and the San Bernardinos of the Transverse Ranges. The Peninsular Ranges are composed primarily of coarse-grained, igneous rocks of the southern California batholith, mostly around 100 m.y. old. The homogeneity of these rocks is expressed in the relatively uniform topographic character of the ridges, peaks, valleys, and gullies in the Jurupas.

Inbound: The Jurupas are on your left going from Colton to the crossing of Sierra Avenue in about five miles.

● At 0.7 mile east of Pepper Avenue, approaching Colton, just to the right is an unusual hill with a flat top and benches indenting its slopes, Slover Mountain (Photo Q–1). It has been in view ahead on the right even before Pepper Avenue. This may be the only major hill in southern California likely to become a hole. The shape is the result of quarrying of carbonate rock (marble) for making cement. Impressive changes in the size and shape of this hill have been observed over past decades by local residents and freeway travelers. If you use this route frequently, watch the hill for changes in configuration.

Inbound: Slover Mountain is on your left as you leave Colton. It is well viewed inbound even before you enter Colton.

● Close to the interchange with I-215, the artificially modified channel of Santa Ana River is crossed and seen briefly to the right. Water usually runs here regardless of the season, even in drought years, although the river bed upstream and downstream may have been bone dry for months. You can thank the San Jacinto fault for this anomalous behavior. The San Jacinto is one of the most active faults in southern California, and its zone of finely ground rock (*gouge*) acts as a ground-water barrier where it crosses the Santa Ana River just left of the I-10 freeway. This underground barrier, known as the Bunker Hill dike, forces water to the surface.

Inbound: The channel of Santa Ana River lies on the right just after you emerge from the I-215 interchange. The San Jacinto fault crosses just a short distance upstream.

● San Jacinto fault passes under the north part of the I-10/I-215 interchange headed northwesterly. It is southern California's most prolific source of historical earthquakes of magnitude 6 or larger, exceeding the San Andreas in that respect. We never pass through this freeway interchange without imploring the earthquake gods to hold everything for a few seconds until we clear the maze of structures.

> ▶ Note your odometer reading near the center of the interchange.

Inbound: Hold your breath inbound as well as outbound.

● Once safely clear of the interchange, you can relax and give more attention to the eastern part of San Bernardino Valley and its bordering features. Due north at 9–10 o'clock on the skyline are the western San Bernardinos and the Lake Arrowhead country, and ahead at 11–12 o'clock is the higher eastern part of the San Bernardinos. South at 2–3 o'clock is the much lower Loma Linda Ridge, part of the Peninsular Ranges, composed of igneous rocks partly buried in much younger deposits of sand, silt, and gravel. Best views are obtained as the freeway rises onto overpasses.

Inbound: These views are seen west from the Highway 30 overpass west of Redlands.

● Even with modest visibility one can see that the eastern San Bernardinos are composed of three east-west ridges, successively higher to the north. San Bernardino Peak and the south land's highest mountain, San Gorgonio Peak (11,499), respectively form the west and east ends of this northernmost ridge. On the north, it is separated from the western San Bernardinos and the Big Bear Lake area by the deep canyon of Santa Ana River. South of this highest ridge, and separated from it by Mill Creek canyon, is the lower Little San Gorgonio Peak (9,140) ridge. Mill Creek fault, regarded as the north branch of the San Andreas, lies along Mill Creek. Next south is a much lower, broader, forelying ridge, separated from the Little San Gorgonio ridge by the south branch of the San Andreas fault. This forelying ridge is bounded

on its south side by the Banning fault. So, as you look slightly left ahead you are seeing a mountain mass sliced into east-west parallel ridges by major faults.

Inbound: These relationships are hard to see traveling west.

● About two miles east of the I-10/I-215 interchange, at California Street, is the turnoff to the San Bernardino County Natural History Museum, invisible amidst orange trees left of the freeway. This is a place well worth visiting; just follow the signs.

① The San Bernardino County Museum is open Tuesday through Saturday, 9 to 5 o'clock, Sunday 11 to 5. It is a first-class operation with outstanding antiquity, archeological, mineral, North American mammal and bird displays, and an unbelievable bird-egg collection. Be sure to take all the up and down ramp-ways. There is much more than just the ground floor. It's a great place for children as well as adults. Admission is free, but contributions are welcome. The gift shop is also worth looking into.

Inbound: The museum is easier to spot and get to on the right. Inbound, watch for signs 1.5 miles west of Highway 30.

● As you leave California Street, San Bernardino Peak looms at 11:45 o'clock, and the western San Bernardinos are good at 9–11 o'clock. At Redlands the character of the nearby landscape changes. The relief of Loma Linda Ridge becomes lower and more subdued, partly because of more burial of the hard crystalline rocks by younger, softer sands and gravels. These deposits are best seen east of Redlands as the freeway passes through a succession of deep road cuts starting beyond Fort Street. The first cuts are wholly within sand and gravel, but beyond the overpass and *"Yucaipa, Right Lane,"* are exceptionally deep, steep-sided road cuts in buried hills of granitic rocks.

Inbound: You enter these deep cuts within a mile beyond Yucaipa Blvd.

● Approaching the rest area about a mile beyond Live Oak Canyon Road, you enter more open country featuring a smooth, gently sloping upland into which broad, steep-walled, flat-floored gullies or arroyos, such as the one at the rest area, have been cut to a maximum depth of about 100 feet. This is rather pleasant terrain. The highway runs across the floor of an arroyo in passing the rest area and climbs up onto the upland at County Line Road near Calimesa, about one mile east. Thereafter, to Beaumont, the highway alternately traverses the upland and drops into arroyos below it. The upland is the top of an alluvial plain built up by streams carrying sand and gravel south from the eastern San Bernardino Mountains.

Inbound: You have been traversing this sort of terrain most of the way from Beaumont.

● For the next few miles, if the day is unusually clear, the Santa Ana Mountains are visible on the far distant south skyline. Santiago Peak (5,687) is the highest point. These mountains lie south of Elsinore fault, a major fracture zone similar to the San Jacinto and San Andreas faults but not as long or as active historically. Some three miles south, to the right near San Timoteo Road, are the low hills of San Timoteo Badlands underlain by the late Pliocene (2 to 3 m.y.) San Timoteo Formation, a land-laid accumulation of sand, gravel, and silt containing fossil remains of various extinct land mammals. Good exposures of these beds are seen along the Riverside-Beaumont Route (Segment R).

Inbound: Santiago Peak is at about 9 o'clock within the first few miles after leaving Beaumont. San Timoteo beds are seen within one to three miles beyond the right-side rest stop.

● As you approach the intersection with Highway 60 (Segment R), San Jacinto Peak (10,801) looms dead ahead on the high skyline. This is a huge mass of crystalline igneous and metamorphic rock, part of the Peninsular Ranges, at the east base of which Palm Springs nestles. Here you pass onto Segment S.

Segment R—Riverside to Beaumont, 24 Miles

● This segment starts at the triple-point intersection of the Pomona Freeway (Highway 60), the Riverside Freeway (Highway 91), and Interstate Freeway 215, near the northeast edge of Riverside (see Figure 4–4).

▶ | Note your odometer reading.

● Your route is southeastward out of the interchange on I-215 toward San Diego. Within 3 miles, at Pennsylvania Avenue, you start up Box Springs Grade and within the next 2 miles rise about 400 feet to emerge on a broad, smooth surface known to geologists as the Perris Upland. Going upgrade are extensive exposures on hill slopes, to both sides, of relatively homogeneous, coarse-grained, granitic, igneous rocks typical of the southern California batholith. You are in the Peninsular Ranges here. The local rock unit is the Bonsall Tonalite, a granitic type of rock richer in calcium, iron, and magnesium than is true granite. On the near left skyline is the western part of Box Springs Mountains, and at Central Avenue their abrupt southwest face is just a mile north. The steepness, relative height, and linear character of this face suggest that it is probably a fault scarp.

Inbound: You drop off the Perris Upland starting down Box Springs Grade beyond the 60/I-215 intersection. Box Springs Mountains make the right skyline, and Peninsular Range granitic rocks are nicely exposed on adjacent hillsides.

• Shortly—1.5 miles—beyond Central Avenue the highway splits. You continue east on Highway 60 toward Beaumont. Beyond the intersection, the openness of the country affords you the opportunity for more distant views. Left is the steep face of Box Springs Mountains; dead ahead on the far distant skyline is Mt. San Jacinto (10,801); and right is the broad, relatively smooth Perris surface, from which isolated steep-sided hills and ridges rise. The smoothness of the Perris surface is the product of weathering and erosion acting upon homogeneous, coarse-grained igneous rocks that disintegrate into grains of uniform size, about like coarse sand and small pebbles. These can be transported on a relatively gentle slope by running water; hence the flatness of the upland. The hills and ridges rising from the plain are erosion residuals not yet consumed. Their height gives a minimum measure of the thickness of rock removed to produce the Perris surface. Geologists consider the evolution of the Perris surface to be more complex than outlined here, but the same principles apply.

Inbound: You see the Perris Upland surface and residual knobs and mountains nicely to the left over the 9 miles from Theodore Street to junction with I-215.

• You proceed eastward across this surface for roughly nine miles before crossing the San Jacinto fault and plunging into the San Timoteo Badlands. Along this stretch, at and east of the Perris Boulevard exit, a prominent ridge of typical granitic rocks is seen to the left, from 9–11 o'clock, 1 to 2 miles distant. Mt. San Gorgonio (11,499), the south land's highest peak, looms on the distant skyline at 10:45 o'-clock. Approaching Nason Street exit, the low ridges of San Timoteo Badlands beds come into view at 10–12 o'clock to the left, about 3 miles away. Even at this distance you should be able to recognize that their topographic characteristics differ from those of granitic knobs and ridges to the south.

② You cross the northwest-trending San Jacinto fault between Theodore Street and Gilman Springs Road that runs along the trace of the fault to the southeast. Beyond are the San Timoteo Badlands (Photo R–1). Three miles right is granitic Mt. Russell, and beyond it, hidden from view, lies Lake Perris, the terminus of the east branch of the California Aqueduct bringing water from northern California.

Inbound: Mt. Russell is at 9 o'clock at Redlands Boulevard. You left the San Jacinto fault and San Timoteo Badlands behind at Theodore Street. The long granitic ridge is on the right beyond Nason Street.

The San Jacinto fault, like its larger cousin the San Andreas, is a right-lateral slip fault, but it is shorter and has a displacement of only about 15 miles compared to the perhaps 10 to 20 times greater offset on the San Andreas. Where crossed by Highway 60, the San Jacinto is a zone about 1 mile wide.

• For nearly four miles beyond Theodore Street the highway traverses the tilted sandstone, siltstone, and occasional conglomerate beds of the San Timoteo Formation, as clearly seen in road cuts and hillside exposures. This is a land-laid deposit of late Pliocene age, perhaps 2 to 3 m.y. old. Complex dissection of these relatively soft but coherent sedimentary beds has created the barren, steep-sided, narrow-ridged topography known as a badland—bad in the sense that it's hard to get around in and only marginally useful for anything. Badlands can be scenically attractive, however, as, for example, Bryce Canyon in Utah. Paleontologists are interested in the San Timoteo beds because they contain the fossilized bones of extinct animals such as camels, ground sloths, deer, and giant land tortoises. The nearby slightly older but similar Mt. Eden beds have an even richer fossil content, including mastodons, dogs, cats, bears, pigs, ground sloths, camels, deer, antelope, raccoons, wolverines, horses, and rhinoceros.

Photo R–1.
View of San Timoteo Badlands off Highway 60 between Gilman Springs Road and Jackrabbit Trail.
(Photo by Helen Z. Knudsen.)

Inbound: At about two miles beyond Jackrabbit Trail and just beyond Call Box 263 you start through the San Timoteo Formation and its badlands, ending at Gilman Springs Road.

● Upon leaving the badlands the highway swings in a broad curve onto the flat floor of upper San Timoteo Canyon, and in about 3 miles rises onto a broad, gently sloping upland surface, the Beaumont Plain, where it joins Freeway I-10 (Segment Q).

Inbound: Leaving Beaumont via Freeway 60 you drive for 1.5 miles across an impressively smooth upland flat, the Beaumont Plain, before starting, at Call Box 293, to drop down into Noble Creek, a headwater tributary of San Timoteo Canyon.

Segment S—Beaumont to Palm Springs, 30 Miles

● This segment begins at the intersection of Highway 60 and Interstate 10 (see Figure 4–4)

and continues eastward through broad, open San Gorgonio Pass, guarded on opposite sides by the two loftiest peaks in the south land, San Gorgonio (11,499) on the left and San Jacinto (10,801) to the right. Elevation of the pass summit, near the southeast corner of Beaumont, is 2,590 feet, so the guardian peaks tower 8,000 to nearly 9,000 feet above it.

▶ | A good place to record your odometer reading.

Inbound: You see these peaks best from farther east and feel you are at the crest of the pass at Cabazon, but the streams say not so.

● Heavily populated coastal areas of southern California, in Los Angeles and Orange counties, are shielded from the harsher climates of the interior by a ring of high mountains created principally by uplifts along faults. These mountains, however, obstruct access by surface transport from the east. The major openings through this barrier for highways, rail lines,

Photo S–1.
Typical woolsack granitic boulders on slope adjacent to Highway 243 to Idyllwild south of Banning.

pipelines, aqueducts, power lines, phone lines, and such are, interestingly, created principally by faults. The three major passes—San Gorgonio, Cajon, and Tejon—come to you largely through the courtesy of the San Andreas fault. San Gorgonio Pass is a good example. It lies along the trace of a plexus of faults within the San Andreas system and is created largely by relative down-dropping of slices along these faults, aided to some extent by erosion of fractured and ground-up rocks within this wide fault zone.

● About two miles east of Beaumont, approaching Highland Springs turnoff, Mt. San Jacinto is prominent on the high skyline at 1 o'clock. Beyond here, especially south of Banning, the hillsides are mantled by large rounded residual boulders (Photo S–1) resembling old-fashioned *woolsacks,* which are so typical of the weathering and erosion of homogeneous granitic rock that they make identification of such rocks easy, even while you are passing at 55 to 65 mph. Boulders become rounded by subsurface weathering (Photo S–2).

Photo S–2.
Core stones in deeply weathered igneous rock, San Diego County, the forerunners of woolsack boulders.

Inbound: These characteristic woolsack boulders are well seen at the mountain foot on the left opposite Banning.

● Near Banning, beyond Sunset Avenue, low, smooth-topped hills a mile to the left at the mountain base are composed of old Quaternary bouldery alluvial gravels (*fanglomerate*), now uplifted by a fault that defines the abrupt, straight, southern edge of this tableland behind Banning. The smooth, gently sloping upper surface of the tableland, part of the Banning Bench, was formed by streams flowing south from the San Bernardino Mountains. These same streams, or their successors, are now dissecting the bench because of the faulting.

Inbound: The south edge of the Banning Bench is seen at 3 o'clock at the west edge of Banning.

● East of Banning, rocks exposed at the mountain foot to the north (9 o'clock) are more heterogeneous. They consist of a complex of gray homogeneous looking fanglomerates and much older crystalline rocks that are more variegated in color, erode to rough craggy, steep slopes, and display a crude irregular banding. Besides having surface boulders, fanglomerates generally make smoother, gentler slopes with rounded ridges except where locally riven by a cluster of small gullies. The crystalline rocks are best seen right at the mountain base about 3.5 miles east of Cabazon. They are largely metamorphics that lie on the north side of the Banning fault, which brought them into contact with the much younger fanglomerates to the south. The fault zone is inclined steeply to the north under the San Bernardino Mountains, which have been lifted up and shoved south in geologically recent time. Some of the fanglomerates involved in the faulting are probably less than 1 m.y. old.

As you leave Banning and can see the mountain front, the fault contact between gray forelying fanglomerate and crystalline rocks runs parallel to the mountain base, about 0.5 mile back in the mountains. Once you train your eye to recognize the difference between fanglomerate and crystalline bedrock, you can pick up this fault contact at many places as you go east to and beyond Cabazon. It is not everywhere the same distance back into the mountains, but you can depend on the highest part of the mountain face being crystalline rock.

Inbound: For 7 miles from east of Cabazon to Banning, watch the mountain front where it is close for views of the fault contact between fanglomerate and crystalline rocks.

③ In this region the San Andreas Fault zone is complex. It splits into several branches that define the successively higher ridges seen within the eastern San Bernardinos. The south branch of the San Andreas slashes obliquely through the mountains to their base about five miles east of Banning, where it either joins the Banning fault or is truncated by it. The fault extending eastward from that point, still called the Banning fault, is regarded as a major component of the San Andreas fault system.

Photo S–3.
Mock-up of a most fearsome carnivorous dinosaur, *Tyrannosaurus rex*, at Cabazon.

● As you approach, you are at Casino Morongo, a bit beyond Hadleys, look right to the mountain base at 2:45 to 3 o'clock. There, below an old road switch-backing up the mountainside, is a large flat-topped embankment of grayish material. This is unweathered rock excavated from a long tunnel on the Metropolitan Water Aqueduct from the Colorado River, which extends southwestward for more than 13 miles under the western spur of Mt. San Jacinto. The capriciousness of nature was demonstrated during construction of this tunnel. Large quantities of water trapped in the rocks seriously delayed completion of the tunnel and bankrupted the contractor. Thus, an aqueduct to bring Colorado River water to southern California was impeded by too much water in rocks along the route.

Inbound: Look for this gray embankment of excavated rock on the left at 9 o'clock opposite Casino Morongo a little beyond Cabazon. It's hard to see in midafternoon light.

● Just 0.5 mile east of the Main Street exit at Cabazon, about 200 yards north of the highway, is a realistic reproduction of two large dinosaurs, a *Brontosaurus,* an herbivorous giant, and a carnivorous *Tyrannosaurus rex* (Photo S–3). This region was never dinosaur country, and these replicas are just for show. Dinosaurs became extinct about 70 to 80 m.y. ago—perhaps just as well, because they would have been a terrible hazard to freeway traffic.

Inbound: The dinosaurs are on the right as you approach Cabazon. If you miss them, the children surely won't. Take the off ramp into Cabazon.

● About 2 miles east of Cabazon, where the freeway describes some curves, the contrast between forelying fanglomerates and craggy, variegated metamorphic rocks north of Banning fault is clearly seen just within the mountain front on the left. In another 1.5 miles, good exposures of metamorphic rock appear right at the mountain base without any of the forelying fanglomerates, which have been removed by erosion.

Inbound: These localities are on the right 1.7 and 4.2 miles, respectively, west of the Highway 111/I-10 intersection.

● At about 11 o'clock as you approach Verbena Avenue is a large mass of uplifted and slightly deformed fanglomerate making up the hill just left of the freeway. The thin brownish zone at the top is an old soil layer.

About 2.5 miles west of Verbena Avenue you can see windmills on the hillside ahead at about 11 o'clock. By the time you get to Verbena Avenue, many more are in sight on hillsides from 11–12:30 o'clock, and there are more to come. Be alert for the turnoff ahead to Palm Springs on Highway 111. You do *not* continue east toward Indio.

▶ | Note your odometer reading at the turnoff. |

Inbound: These exposures of fanglomerates are on the right as you rise on the overpass of Highway 111 over I-10.

● After the turnoff on Highway 111 you see the sandy floor of San Gorgonio wash to the right, backed by hill slopes of granitic rock. Ahead deposits of wind-blown sand are banked against and partly mantle spurs projecting north from the base of San Jacinto Mountain. This sand is picked up from San Gorgonio wash by the frequent strong winds that blow eastward out of San Gorgonio Pass into Coachella Valley.

Inbound: Within the last 2.5 miles approaching the 111/I-10 junction Whitewater bouldery wash is to the right and San Gorgonio wash is on the left.

● In passing *"Whitewater, 3 miles,"* on the right, look left at about 10 o'clock for a near skyline view of a rounded, dome-shaped hill with communication towers, a small white building near its summit, and many windmills on its flanks. That is Whitewater Hill, just east of the mouth of Whitewater Creek, which heads high on the south slope of the San Gorgonio Mountain. This hill is unusual because its shape is determined by deformation of such recent date that the slopes of the dome are only slightly modified by radial gullies, although its southern part has been cut away by Whitewater Creek. Not many places in the United States have landforms created by deformation that exist essentially in their pristine condition.

Inbound: Whitewater hill is about at 3:30 o'clock as you round Windy Point and remains in view as you continue west.

● As you pass *"Whitewater, 3 miles,"* Windy Point is in view dead ahead. Perennial Palm Springs travelers can testify that it is well named, and both sides of the point are mantled by wind-blown sand (Photo S–4) derived from the sandy floor of San Gorgonio wash. Windy Point is distinguished by the strong layering and foliation visible within metamorphic rocks that compose this spur. This banding is seen dead ahead approaching and crossing the bridge over San Gorgonio wash and in the road cut beyond (Photo S–5). You will seldom see better metamorphic banding.

The Windy Point exposures encompass a wide variety of rocks, including beds of limestone now largely metamorphosed to marble (Photo S–6). This sequence is so distinctive that boulders derived from this and similar exposures nearby are easily recognized within some of the coarse fanglomerate deposits far away on the opposite side of the valley, for example, at Garnet Hill (Figure 4–4), north of Palm Springs.

Inbound: Banded metamorphic rocks are in good view on the left as you pass Windy Point before the San Gorgonio wash bridge. The lee (near) side of Windy Point has a mantle of wind-blown sand, a falling dune, used by over-terrain vehicles. You also see sand on the windward side (Photo S–4).

● The wide alluvial plain making up the valley floor left of the highway at and beyond Windy Point is the product of floods generated within

Photo S–4.
Accumulation of wind-blown sand and silt on windward side of Windy Point on Highway 111 to Palm Springs. Trails made by over-terrain vehicles. (Photo by Helen Z. Knudsen.)

Photo S–5.
Strongly foliated and banded metamorphic rocks in road cut at Windy Point, Highway 111 to Palm Springs.

Photo S–6.
Carbonate (marble) layer in banded metamorphic rocks of Windy Point. (Photo by Helen Z. Knudsen.)

the San Gorgonio wash and Whitewater Creek drainages, with Whitewater Creek probably dominant. It is from this plain, largely between here and Indian Avenue, that most of the wind-blown sand is derived which so inconveniences residents of Coachella Valley farther east. Observations made over a 15-year period in an experimental wind plot about 0.5 mile west of Indian Avenue showed a direct correlation between the amount of wind-driven sand moving past the station and the amount of flooding that occurred each year on the alluvial plain left of the highway. Each flood renews the supply of sand available to the wind. Curiously, long-range control of wind-blown sand in western Coachella Valley may depend more on

flood control than on sand control. Wind-blown sand and silt are remarkably effective erosive agents (Photos S–7 and S–8).

Inbound: You have been driving alongside this alluvial plain from Chino Canyon to Windy Point. Between Chino Canyon fan and the first rock spurs, Mt. San Gorgonio is dead ahead on the skyline.

● Within a mile beyond Overture Drive and the Palm Oasis settlement, you pass around two more rock spurs with well-laminated rock. About 0.5 mile beyond the second spur, at *"Palm Springs City Limit"* to the right at the

Photo S–7.
Wind-blasted boulder of hard
crystalline rock, a ventifact, on Garnet
Hill. Pitting is typical of faces steeply
into wind.

Photo S–8.
Wind-blasted common red brick from Whitewater alluvial plain, after six years exposure.

Photo S–9.
Protruding lobe of rocky debris at mountain base, in foreground, is product of an old rock slide.
Locale is between Palm Oasis and Chino Canyon fan. (Photo by Helen Z. Knudsen.)

mountain base is the large, projecting lobe of an old rock-fall slide (Photo S–9). It is about 150 feet high at the toe, projects several hundred feet out onto the valley floor, and consists of large rock blocks.

Inbound: This slide lobe is even better seen on the left within 0.5 mile beyond the large flood-control dike on Chino Canyon fan.

● Also from here at about 9–9:30 o'clock far across the valley, is the settlement of Desert Hot Springs nestled against the base of the Little San Bernardino Mountains. Desert Hot Springs lies on the trace of Mill/Mission Creek fault, the northern strand of the San Andreas system. Fractures within this fault zone provide passageways to the

surface for waters that have descended deeply enough into the earth's crust to be heated as high as 200° F, although averaging 122° F.

Movement on Mission Creek fault caused the 1948 Desert Hot Springs earthquake, a sharp shock of 6.3 magnitude that did some local damage. The succession of smaller shocks following the 1948 quake continued into 1957 and was one of the first aftershock sequences to be thoroughly recorded and studied by modern instruments and techniques. Because Mission Creek fault is inclined northward and the *focus* (point beneath the ground where the quake movement originated) was at a depth of about 10 miles, the *epicenter* (point on the surface

Photo S–10.
Head of bouldery Chino
Canyon fan with Mt. San
Jacinto in background.
Chino Creek must be
diverted here and
channeled down west side
of fan to protect valuable
properties downslope.
(Photo by Helen Z.
Knudsen.)

directly above the focus) was well to the north
within the Little San Bernardino Mountains
rather than in the middle of Desert Hot Springs.

*Inbound: Desert Hot Springs is seen at 3–3:30
o'clock across the valley from the apex and
western part of Chino Canyon fan. Wind farms
are abundant on the foreground alluvial plain
of the valley floor.*

● A half mile beyond *"Palm Springs City
Limit"* you get a good view of the boulder-rich
(Photo S–10), well-formed Chino Canyon fan
with a huge flood-control dike on its western
side that diverts the channel of Chino Creek.
Because alluvial fans are built by streams that
swing back and forth like fire hoses, anyone living on a fan surface can expect trouble from
floods, unless for some reason the stream has
started to dissect the fan and now flows in a
confined course. At considerable expense the
stream on Chino Canyon fan has been artificially confined to the west flank to protect the
extensive real estate developments on the central and eastern parts. People living on alluvial
fans should know how fans are built, what their
stage of development is, and whether they are
being dissected. Tramway Road takes off up
Chino Canyon fan to the lower end of the aerial
tramway 0.9 mile from the flood-control dike.
Continuing, you come to Racquet Club Road
intersection well into the densely built area of
Palm Springs.

*Inbound: The heavily built area between Rac-
quet Club Road and Tramway Road is on the
surface of the Chino Canyon fan and would be
at risk if it were not for the flood-control dike
to the west.*

Photo S–11.
Typical windmill farm on floor of Coachella Valley.

● Residents of Palm Springs may not fully recognize and appreciate the remarkable wind shadow (becalmed area) in which their city lies. Winds blow westward with strong velocity and great frequency through San Gorgonio Pass into Coachella Valley. Thus, more often than not, travelers to Palm Springs from the west experience strong winds through the pass, but upon arrival in Palm Springs find almost a dead calm. The huge mass of Mt. San Jacinto protects the city. In addition, the mountain, being lofty, also captures a lot of rain and snow—a principal source for the water that makes Palm Springs a garden spot in an otherwise dry desert region.

Inbound: Travelers leaving Palm Springs are often surprised to discover they have to drive home into the teeth of a strong west wind.

The elevation of Palm Springs is about 500 feet, and the top of Mt. San Jacinto is at 10,801 feet. This results in an abrupt vertical relief of more than 10,000 feet on the steep eastern face of the mountain, which is some-

what greater than the vertical relief along the highest parts of the eastern face of the Sierra Nevada in the Mt. Williamson-Mt. Whitney area. It also exceeds the vertical relief in most other parts of the contiguous United States except Death Valley.

④ If you find the abundance of windmills (Photo S–11) on parts of the floor of Coachella Valley and on surrounding slopes distasteful, be comforted by the knowledge that you have a lot of company within the local populace. At one time the city of Palm Springs brought suit against the developers of the wind farms and the Bureau of Land Management for violating their environmental impact agreements by installing more windmills than permitted. There are currently about 4,000 in operation. The suit was settled out of court.

Generating power from the wind is an attractive concept, particularly if the wind farms are established in remote, unpopulated areas. Even so, it has some drawbacks. Detached flying propeller blades are dangerous, probably capable of cutting through a passing car as though it were a chunk of swiss cheese. A forest of operating windmills creates an eerie pervasive noise, kills birds, and possibly distorts television reception. The mills themselves disfigure the landscape. The power generated has to be used immediately; there is as yet no satisfactory mode of storage. Wind-generated power is more expensive than power generated by conventional means because of the large capital investment required. The mills and generators require careful maintenance and, unfortunately,

many of the sponsors of the limited partnership consortia that installed the wind farms have moved on to other endeavors. Wind farms should probably not be erected in urbanized areas such as Coachella Valley, no matter how favorable the wind regime.

● The Coachella Valley area around Palm Springs is full of interesting geological features worth a booklet in their own right. Regretfully, they cannot be treated here. See California Division of Mines and Geology Special Publication 94, by Richard J. Proctor, 1968, for details on some of them.

SAN FERNANDO PASS TO SAN JOAQUIN VALLEY

Segment T—San Fernando Pass to San Joaquin Valley, 59 Miles

● This segment follows the Golden State Freeway (I-5) from the northwest corner of San Fernando Valley to the southwest corner of San Joaquin Valley (Figure 4–5). In windy weather, visibility at the west end of San Fernando Valley can be great, so look around if it's that kind of day. The Santa Monica Mountains make the distant south skyline, the rough and rugged Simi Hills close the west end of the valley, the higher but smoother Santa Susana Mountains lie to the northwest, and the steep western San Gabriels loom to the north. The contrasting appearance of these mountain masses reflects differences in the geological structures and rocks composing them. Rocks in the San Gabriels are the oldest, hardest, and most complex; the Santa Monicas, Simi Hills, and Santa Susanas then follow in order of decreasing rock age. San Fernando Pass is an historical escape route through this encircling barrier. Geological relationships in the pass are briefly described in the first paragraphs of Segment J.

Inbound: All travelers must be four to six miles beyond San Fernando Pass to see these larger relationships.

● At the interchange of I-5 and Highway 14 (Antelope Valley Freeway).

▶ Note your odometer reading.

Observe rock exposures in the large road cuts and in hillside bluffs around the interchange. As you pass on I-5 under the maze of arches of this "bucket of worms interchange" (see Photo J–1) a quick look right gives a glimpse of a large road cut on the Antelope Valley Freeway

with a lovely *syncline*. The rocks here are sea floor deposits of Mio-Pliocene age (4 to 8 m.y.) in the upper part of the Towsley Formation.

Inbound: It is difficult to see much geology at the interchange inbound until you are beyond it, where large road cuts expose severely deformed sandstone, shale, and conglomerate beds.

● Upon clearing the interchange you are driving up Weldon Canyon, but you don't see much of interest, except for steeply inclined beds of sandstone and shale on the left, until you pass over the divide into Gavin Canyon beyond the big truck parking area, a mile from the center of the interchange. Then hillside exposures of sandstone and conglomerate beds with various degrees and directions of inclination, *dip,* are seen, especially on the right. From the crest of the divide to Calgrove Boulevard the steep, north-facing slopes left of the freeway have a relatively dense, green vegetative cover with many oak trees and clumps of conifers (big-cone spruce) that contrasts sharply with brush-covered slopes on the right that face south.

Farther down the canyon, on the right at 9 o'clock behind Call Box 474, the bluffs of sandstone are even more prominent and contain much more conglomerate. They are part of the younger Pico Formation (2 to 5 m.y.), also of marine origin. From Call Box 474 essentially all the way to Calgrove Boulevard the skyline at 11:30–11 o'clock displays near-vertical bedding. These steeply inclined beds are in the core of the tightly folded Pico Anticline, an up-bowed fold and one of the earliest significant oil-producing structures in southern California.

Inbound: These relationships are seen ascending toward the head of Gavin Canyon from Calgrove Boulevard, but seeing the Pico Anticline is virtually impossible without stopping, which is not advisable on this or any freeway.

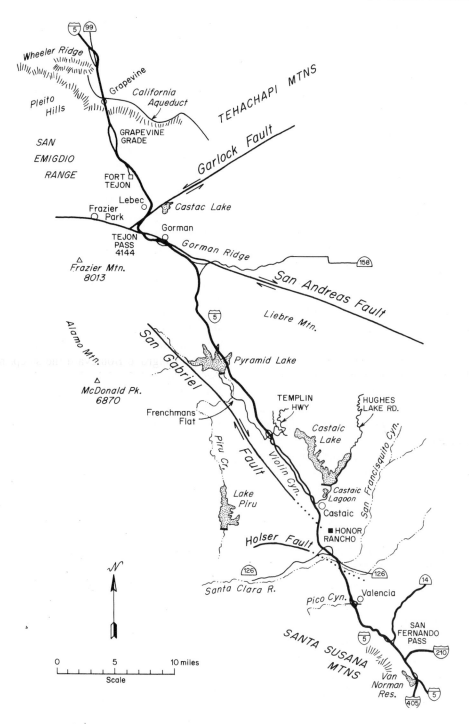

Figure 4–5.
Segment T, San Fernando Pass to San Joaquin Valley.

● At *"Calgrove Blvd. 1/2 Mile"* directly ahead just left of the freeway is an exposure of strata dipping about 35° north. They are well out on the north limb of Pico Anticline. The steep beds near its core can still be seen at 10 o'clock from near Call Box 486.

Inbound: These exposures are too close to the freeway to be easily seen going in.

● You soon pass the Lyon Avenue-Pico Canyon Road exit and overpass, and beyond at *"McBean Parkway, 3/4 Mile,"* can again see, now at about 9 o'clock, the steep beds at the anticlinal core. The Valencia golf course is just barely visible through the trees along here on the right. It occupies an alluvial flat built by Pico Creek.

Shortly you top a divide, where McBean Parkway turns off, and in 0.2 mile come to an overpass with *"Valencia Blvd. 1 Mile"* on it. Thereafter you go downhill along a stream course that was diverted by natural capture from the right, just beyond the overpass at Valencia Boulevard with *"Saugus 1 Mile"* on it. Vegetation now screens the course of the capturing stream, but you follow the old, abandoned valley all the way down to Magic Mountain Parkway. The little narrow golf course to the right also uses this abandoned stream valley.

Stream capture is a relatively common occurrence. What happened here is that a small tributary to the Santa Clara River worked headward about at right angles to the present freeway course. It breached a divide and discovered our freeway stream, also a tributary of the Santa Clara River-making junction farther downstream. The little captor tapped our freeway stream on the shoulder and said, "Hey, pal, follow me, and I'll show you a shorter route to the Santa Clara." Because our freeway stream wanted to go to the Santa Clara as quickly and easily as possible, it accepted the invitation. Once having turned and followed the course of

the capturing stream it was committed to that route, leaving its abandoned bed as a suitable site for the golf course and freeway.

Inbound: You drive up the abandoned stream valley between Magic Mountain Parkway and the Valencia Boulevard overpass.

● About 0.3 mile beyond Magic Mountain Parkway turnoff, is the bridge across the Santa Clara River. Neither bridge nor river is easily recognized. The bridge starts 0.1 mile beyond the overpass above Magic Mountain Parkway and ends 0.2 mile farther. The tree-filled course of the river is crossed in the process. Look for a solid cylindrical pipe-like railing to identify the bridge. The Santa Clara drains a large area on the northwest flank of San Gabriel Mountains, and it occasionally works up a pretty good flood. In the olden days, this and other bridges downstream, all wooden, were episodically destroyed by such floods. Humans sometimes inadvertently give nature a helping hand in this endeavor. In 1928 a huge flood swept through here, removing the bridge and cleaning out cottonwood and willow trees much larger than those now growing along the channel.

That was the disastrous St. Francis (or San Francisquito) flood. After completing the Owens Valley Aqueduct in 1913, the City of Los Angeles constructed storage reservoirs southwest of the San Andreas fault, which the aqueduct crosses in an underground tunnel, partly as a safety measure and partly as a means of storing excess water. One such reservoir was built in San Francisquito Canyon, a tributary of the Santa Clara on the north side about 0.5 mile east of here. In those days not nearly as much attention was paid to geological aspects of dam sites as is given today, and the foundation for the San Francisquito Dam proved to be geologically unsound. As a result, the dam collapsed at midnight on March 12, 1928, sending a wall of water, initially more than 100 feet high, rushing down San Francisquito Canyon and the

Santa Clara River into the sea between Port Hueneme and Ventura. Damage was extensive and loss of life was considerable; just how great will never be known, because a favorite hangout for hoboes was in the large cottonwood trees along Santa Clara River, and no one kept a census on the hobo population.

Inbound: Scars of the flood are no longer visible here, but just imagine the present bridge and trees along the stream channel completely scoured away, and you get a sense of the scene.

● About 100 yards right of the freeway, 0.1 mile beyond the Santa Clara River, is an abandoned gravel excavation that has left a cut bank at the end of a low ridge. The materials excavated were sandy gravels of the Saugus Formation, here turned up into nearly vertical position by drag along the Holser fault that crosses the freeway obliquely. When the quarry was fresh, near-vertical bedding in the excavation was easily seen. Now, because of vegetation, erosion, and weathering, the bedding is hard to see except at the top of the cut.

Inbound: Vertical beds are better seen inbound beyond the truck inspection station approaching Magic Mountain Parkway.

● As you approach the truck inspection station, look to the low hills about a mile beyond at 12–1 o'clock to see the varied topographic forms developed by erosion on tilted Saugus beds.

Inbound: This extensive exposure of Saugus beds is on the left after you go under the overpass of Highway 126.

● Beyond the overpass for Highway 126, to Ventura, if the day is clear, rounded Cobblestone Mountain looms on the far distant skyline dead ahead. This is a largely crystalline rock mass in the back country of Ventura County.

Inbound: Cobblestone is not easily viewed inbound.

● A little less than a mile after you pass Highway 126, you cross Castaic Creek (usually dry), followed shortly thereafter by the Hasley Road turnoff. Beyond that are the green fields of Wayside Honor Rancho right of the freeway, best seen at 1–3 o'clock opposite Call Box 566. This is a Los Angeles County detention camp, probably the only one of its kind in the United States—perhaps the world—that once had its own oil field.

Inbound: Wayside Honor Rancho is on the left across the bed of Castaic Creek south of the elevated pipelines opposite Call Box 583.

● Near the north edge of Honor Rancho you cross the now-inactive San Gabriel fault, buried beneath valley-bottom alluvium and the Saugus beds of the nearby hills. The fault crossing is about opposite the elevated light-colored pipeline seen a little right of the freeway where it comes out of the ground to cross the bed of Castaic Creek. This is near *"Parker Road Castaic 1 Mile."* The San Gabriel is probably an early member of the San Andreas system. It died when the present San Andreas zone developed, which is seen ahead near Gorman. From here the fault and freeway take up nearly parallel courses, just a few miles apart, extending northwestward together for nearly 20 miles. In its day, the San Gabriel was a real humdinger. Differences in bedrock units on opposite sides and the composition of stones within coarse deposits along its trace suggest a major right-lateral displacement of at least 18 miles.

Inbound: The elevated pipeline crossing Castaic Creek is easy to spot. It is opposite Call Box 583. Wayside Honor Rancho buildings are far left from Call Box 577.

Unfortunately, you don't see the trace of the San Gabriel fault from the freeway, but spectacular deposits with large angular fragments of crystalline rocks, the Violin breccia derived from its scarp, lie a little west of the freeway route and are world-famous among geologists. For more information on the San Gabriel fault in the mountains behind Pasadena, *see* Segment I.

Photo T–1.
Angular unconformity between gently tilted and faulted Pleistocene Saugus beds and still younger overlying fluvial deposits, largely sand and gravel.

● About 0.3 mile beyond *"Lake Hughes Road, Second Right,"* the face of a steep cut about 100 feet left of the freeway exposes a sharp angular unconformity between gently tilted and faulted sandstone and shale beds and overlying brown, oxidized stream gravels (Photo T–1). It is hard to see outbound, but the inbound view is good. Any angular unconformity tells a story. The tale here involves deposition of sandstone and shale, tilting and faulting, erosion, renewed deposition (of the stream gravels), followed by further uplift and erosion to expose the whole sequence.

Inbound: This cut is on the right 0.5 mile beyond the overpass of Parker Road.

● **Special Note:** Anyone wishing to view this unconformity more closely can do so from the west-side frontage road (The Old Road); outbound by turning off at Hasley Road and going north or inbound by turning off at Parker Road and going south. The exposure is 1.9 miles north (0.9 mile north of Hill Crest Park-Evergreen Subdivision), and 0.5 mile south from Parker Road. An iron picket fence and gate mark the easiest access.

● An interesting side trip is available at Castaic for those who have spare time. Turn off on Lake Hughes Road and follow signs to the Lake Castaic visitor's center, a distance of only three miles. En route you get a good look at the huge earth-fill Castaic dam, which has earlier been in view from I-5. The body of water seen below the dam is an *after bay,* Castaic Lagoon, which is bordered by spacious picnic grounds, a good place to eat lunch. Beyond Castaic Lagoon the road ascends, initially through sandy gravels of the Saugus Formation (1 m.y.), and then through shales, mudstones, siltstones, and fine sandstones of the Charlie Canyon or Castaic Formation of upper Miocene age (roughly 6 to 10 m.y.) and marine origin. Coming down you will see frequent changes in dip of bedding in the Castaic sediments showing that they have been strongly deformed. The disheveled topography of the hill slopes also suggests considerable sliding, slumping, flow, and creep within these rocks. Landslides were a problem during excavation of the east abutment for the dam.

Across Castaic Lagoon to the west are the excavations, yards, and buildings of Castaic Brick Company, a major operation that uses the fine clay-rich shales of the Castaic Formation for manufacture of its products. This site is best viewed coming down from the visitor's center.

Signs leading to the visitor's center can be a little misleading as you ascend the grade. Ignore all paved roads turning left. Just take the first paved road to the right and follow it to the modern building atop the hill. Parking is ample, restrooms good, displays educational, and the overlook of Lake Castaic and its surroundings impressive.

● Lake Castaic is an extensive body of water, with a west arm up Castaic Creek and an east arm up Elizabeth Lake Canyon. On its far side the consistent dips in Castaic Formation beds contrast with the more complex structure exposed along the road just traversed. The entire route of the California Water Project and all dam sites were carefully studied by a large corp of engineering geologists to insure, as far as possible, against any San Francisquito-like catastrophe. The water in Lake Castaic exceeds that stored in the San Francisquito reservoir.

Inbound: To some degree any large body of impounded water is a potential danger for areas downstream, no matter how stable its dam.

● As you return to I-5, better views are obtained of the Castaic Formation beds in road cuts descending from the visitor's center, and the change to gravelly Saugus beds in the last 200 to 300 yards should be more apparent. In the middle of Castaic village, the old two-lane concrete road taking off right is part of the famous (or infamous) original "Ridge Route" (see street sign) from the Los Angeles Basin to San Joaquin Valley. It was certainly one of the most tortuous, winding, California highways of its day—the 1920s and 1930s.

Inbound: The preceding paragraph is inbound in orientation.

● If you are interested in laying hands on a modest exposure of Violin breccia close to the San Gabriel fault, when you get back to the freeway just continue straight ahead under it to a boulevard stop at The Old Road. Turn right there and follow The Old Road almost to its end, about 0.8 mile. There turn left onto Victoria Road and go to its west end, park on the left side. The dark-brown bluff there is Violin breccia. You can see it without descending into the gully beyond the street's end, but exposures down there are more extensive (Photo T–2). Then retrace your course to gain the freeway northbound.

The breccia is a most unusual geological formation, not only because of its constitution but also because of its geometrical configuration. It forms only a narrow strip along the northeast side of the San Gabriel fault for a fair number of miles northwest from here. Although most formations are sheet-like, the Violin breccia is long, narrow, and thick, a curious shape for a formation. The reason is that it is made up of the coarse rock debris that was shed into the Ridge Basin from the actively rising scarp of the San Gabriel fault, by rock slides, debris flows and streams.

Outbound travelers who did not go to the visitor's center can exit I-5 at the Lake Hughes turn off, go left under the freeway to The Old Road and follow directions.

Photo T–2.
Violin breccia outcrop in canyon wall just beyond west end of Victoria Road. Large clast in lower left, 4–5 feet diameter.

Inbound: The violin breccia bluff is visible up Violin Canyon on the right from the outer lane of the freeway at Call Box 607. To see it, exit at Lake Hughes Road, turn right at first boulevard stop, and you are on The Old Road.

▶ | Record your odometer reading at the on-ramp to I-5.

● Back on the freeway, you ascend the floor of Marple Canyon. You may be surprised to discover that the southbound freeway lanes have switched position and are now on your right side and well up on the east wall of the canyon, high above. At Call Box 608 is a good place to note this relationship. It was easy to miss this switch, because the other lanes passed over you back near the Lake Hughes Road complex. They get back in their proper position, on the left, as you pass over them near the top of the

grade just before *"Templin Highway, Exit 1 mile,"* in about 6 or 7 miles. Now why do you think the highway engineers arranged things this way?

Inbound: These relationships are more clearly seen and appreciated inbound going downhill from "Elevation 2,000 ft." At Call Box 613 the other freeway lanes are on the canyon floor far below you. By now you have probably realized the switching of freeway lanes was done to give inbound trucks the gentlest downgrade possible.

● The smooth, grassy slopes adjoining the canyon floor you are ascending are underlain largely by shales and mudstones of the Castaic Formation. In 0.8 mile, opposite the large *"Turn off air conditioners"* sign, a major tributary, Violin Canyon, comes in from the left. The dark-brown bluff at the base of the hills about a quarter mile left at 9 o'clock there will catch your eye. That is the exposure of the famous Violin breccia, visited by those who ventured the detour via The Old Road to the end of Victoria Road.

● You are about to enter and travel across a most unusual accumulation of sedimentary beds: the Ridge Basin Group. At their base they interfinger with layers of the upper Miocene Castaic Formation through which you are traveling. The Castaic Formation was deposited in the ocean, but the Ridge Basin sediments were laid down in a land-locked basin bordering the sea. They are unusual in their great thickness— more than 30,000 feet—and localization within a relatively small narrow basin. They are surprisingly uniform, well bedded, and fine-grained, except for the Violin breccia, considering their terrestrial origin. The beds do get coarser and less uniform near the top of the sequence in Hungry and Peace valleys, near Gorman, but that is still 25 miles ahead. They are also somewhat coarser toward the east and west margins of the basin.

● Beyond Call Box 612 you are climbing up out of your canyon, and soon you see on the left steeply inclined beds, presumably deformed by drag along the San Gabriel fault. At Call Box 626 note that you are almost up to the level of the inbound freeway lanes, still on your right.

About 0.5 mile beyond *"Elevation 2,000 Ft."* brown beds associated with lenses of Violin breccia are seen about 0.5 mile left of the freeway. The deep road cuts you pass through near the top of the grade expose dark shales and gray-to-brown sandstone layers of the Castaic Formation, but the farther north you go the closer you come to the shoreline of the old Castaic sea. Road cuts beyond *"Angeles National Forest"* expose sandstones and shales that are probably part of the transition zone between marine Castaic beds and terrestrial sediments of the Ridge Basin Group. Between *"Templin Highway, Exit 1 Mile"* and the actual exit, out-

croppings of resistant sandstone are more apparent on the hillsides. This is an expression of the change from marine to terrestrial deposits.

Inbound: You traverse this transition zone from terrestrial beds with many thick sandstone layers to marine shales beyond Templin Highway but cannot see details of features viewed along the outbound lanes.

● At and beyond *"Templin Highway Right,"* the dark beds exposed in road cuts are Ridge Basin shales, accompanied by brown and gray sandstone layers. The white material seen in road cuts is a secondary encrustation of alkali salts left on the surface of fractures by evaporation of seepage waters.

Inbound: These exposures are seen as you approach Templin Highway, mostly to the left.

● Just before the Templin Highway separation, Violin Summit (2,591) is reached.

> ▶ Note your odometer reading at Templin Highway crossing.

The crest of the freeway grade is attained about 1.5 miles farther along. There, near Call Box 676, you get a first good look (northwestward) out over the huge dissected pile of sedimentary beds filling the Ridge Basin. This area is drained by Piru Creek and its many branches. The old freeway that followed the bottom of Piru Canyon has been visible on the left since Templin Highway. Construction of Pyramid Dam and Reservoir, which you'll be able to spot shortly ahead, required relocation of the freeway onto its current higher course. The new location gives motorists the opportunity to look west into the high rugged country along the Los Angeles-Ventura counties line. The deep gorges and canyons seen to the left, 4 to 5 miles beyond Templin Highway, lie largely in the Piru Creek drainage, as do the cluster of the higher peaks, including Alamo Mountain,

Photo T–3.
Shale-rich (dark) phase of Ridge Basin deposits in canyon of Piru Creek below Pyramid Lake dam.

sometimes snow-capped in winter. They are composed of crystalline rocks, much older than the 6 to 8 m.y. beds of the Ridge Basin.

Exposures of Ridge Basin beds in deep road cuts on both sides were excellent when the cuts were fresh, but weathering and vegetation have taken their toll. Dark layers are mostly shale and mudstone (Photo T–3), lighter beds are largely sandstones (Photo T–4), laid down by streams or sublacustrine currents (Photo T–5). Road cut exposures improve northward and sandstone beds are usually obvious. Inclined ledges on hillsides mark the traces of particularly resistant sandstone layers. These beds dip northwestward in the direction you are traveling, but a few miles to the west, toward the higher rugged peaks, the beds dip east and farther east, they dip west. You are traveling roughly up the axis of a large down folded area, a syncline.

Inbound: Views into the high rough country to the west are good at and beyond the new Pyramid Lake overlook, and the old freeway is seen in places. Exposures of Ridge Basin beds in road cuts and on hill-slopes are extensive for many miles.

● From 4 to 6 miles beyond Templin Highway keep looking westward for views into the rough high country dissected by the deep gorges of Piru Creek and its tributaries. Frenchman Flat on Piru Creek can still be reached by way of the abandoned freeway, and one of the best places to see Violin breccia is to walk down Piru Creek about 0.5 mile from that flat. The old freeway is accessed by turning off on Templin Highway, going left under the current freeway and then right on the old highway. It's about 5 miles to the flat. There, truly spectacular exposures of the breccia, with

Photo T–4.
Sandstone ledges within
Ridge Basin deposits, Piru
Creek, below Pyramid Lake
dam.

huge angular blocks of crystalline rock, make up the walls of Piru gorge, and just beyond is the San Gabriel fault.

Inbound: To visit Frenchman Flat turn right on Templin Highway and go about 5 miles northwest (right) on the old freeway.

● As you drive mile after mile northwesterly through the Ridge Basin sedimentary beds, you begin to appreciate their aggregate thickness— approaching 33,000 feet. That's a lot of mud and dirt! The fine shale beds (mostly dark) are thought to be lake deposits and river floodplain accumulations. The lighter sandstone layers are probably stream-bed deposits.

Inbound: You traverse these beds between "Pyramid Lake Next Right" and Templin Highway.

● Between six and eight miles beyond Templin Highway, first at Call Box 678 and especially at Call Box 722, the large mountain looming on the distant skyline nearly dead ahead is Frazier Mountain, a geologically fascinating feature. Keep observing it as you get closer. At eight miles beyond Templin, near Call Box 736, the waters of Pyramid Lake appear on your left. By looking back to about 7:30 o'clock, glimpses can be caught of Pyramid Dam. Southbound travelers see it more easily.

Inbound: Frazier Mountain is behind you, but Pyramid Dam and Lake can be seen from the new overlook and are in view to the right a little beyond Vista del Lago Road.

● A major visitor's center was under construction in 1993 by the California Department of Water Resources, on the left about eight miles from Templin Highway, in view opposite Call Box 742. Access from outbound lanes will be furnished by a new overpass. This overlook will provide a good view of the eastern part of Pyramid Lake and of the dam. Presumably nice facilities and descriptions of the functions of the California Aqueduct and Pyramid Lake will

Photo T–5.
Ripple marks on bedding plane of sandstone layer in Ridge Basin deposits.

be provided along with ample parking. The name, Pyramid, comes from a striking landform in Piru Gorge, now submerged, which was suggestive of Egyptian pyramids.

● You can think of Pyramid Lake as a gigantic storage battery; its principal function is power storage. From its east side a tunnel leads seven miles southeast to the edge of Castaic Canyon near the east end of Templin Highway. There the water is fed under a near-thousand-foot head to the turbines of Castaic Powerhouse operated by the Los Angeles Bureau of Water and Power.

This installation can be visited by turning east on Templin Highway at the Castaic Power-

house sign and following the paved highway about 2 miles to a vista point, with good exposures of Ridge Basin beds along the way. There you can see the powerhouse, perhaps hear it hum, and note on the canyon floor downstream a modest body of water called a *forebay,* which is separated from the head of Lake Castaic by a dam. The water in the forebay will be pumped back up to Pyramid Lake late at night or over the weekend for storage until it is needed again. Clearly no such system is 100 percent efficient, so power is lost in this process. Nonetheless, it is financially feasible because of the higher electrical rates charged at times of peak demand when the water is run back down through the powerhouse. One would think the water must get worn out after recycling through the system several times. However, new water from the aqueduct is continually pouring into Pyramid Lake and some through flow into Castaic Lake means that no particle of water is locked forever into this recycling system.

Inbound: An expansive parking area and large building are being constructed on the right 0.3 mile beyond the overpass with a sign to the Pyramid Lake overlook via Vista del Lago Road, which gives a fine view of the lake and dam. Castaic powerhouse is viewed via the Templin Highway turnoff ahead.

● About 10 miles beyond Templin Highway, near Call Box 756, the upper Pliocene to Pleistocene Hungry Valley beds (1 to 3 m.y.) composing the upper part of the Ridge Basin Group, come into view. They are coarser, lighter in color, less well bedded, and more weakly consolidated and look largely nonlacustrine. Here, the country opens out, and the character of the surrounding slopes changes. In less than 2 miles you are down onto the flat alluvium filled floor of Cañada de los Alamos, and Los Alamos barranca is seen sharply cut into the alluvium as it extends westward toward Hungry Valley at about 9 o'clock. This is just short of *"Pyramid Lake, Next Right."*

Inbound: Hungry Valley beds extend to about where the highway starts to climb up from the smooth, flat, alluviated Cañada de los Alamos floor.

> ▶ Note your odometer reading at the Hungry Valley Road turnoff.

● Here are good views of the Hungry Valley beds and another look up Los Alamos barranca. The large cottonwood and willow trees growing on the barranca floor suggest it has a modest antiquity, but even so, this dissection was probably a recent event in geological (if not historical) terms.

Inbound: Views of Los Alamos barranca are to the right approaching Hungry Valley Road.

● The route continues north along the east side of the smooth, alluviated floor of Gorman Creek, a Los Alamos tributary. Good exposures of Hungry Valley beds are seen on hill-slopes to the left and right.

Inbound: Good exposures of Hungry Valley beds lie to right. The old canal that carried aqueduct water into Pyramid Lake is just right of the highway in vicinity of Call Box 807.

● About 2 miles beyond Hungry Valley Road turnoff, near Call Box 794, you see a quarry ahead at about 11:50 o'clock a few hundred yards left of the freeway, and you pass it in less than a mile. Your geological eye should recognize that the gray rock in the quarry differs in appearance from the surrounding Hungry Valley sedimentary beds. It is a mass of older, badly fractured, coarse-grained granitic igneous rock, which goes by the special name of *quartz monzonite*. This body of igneous rock is unusual in having been squeezed, in a solid state, into the core of an anticlinal fold. That's why it is so badly fractured and so easily quarried. On top of the quartz monzonite are about 50 feet of near-horizontal brownish gravel beds, younger than the deformation and the Hungry Valley Formation. The quarry is best seen from *"Lancaster, Palmdale, Highway 138, 1 mile."* About 0.25 mile before passing the quarry the freeway crosses the concrete canal that formerly carried water down Gorman Creek to Pyramid Lake.

Inbound: Going south you have to be beyond the quarry to recognize it. Look back at 4 o'clock from Call Box 809.

● Beyond the quarry, road cuts, especially left of the freeway, expose steeply dipping, deformed Hungry Valley beds. Ahead on the skyline is Gorman Ridge. Its smooth, grassy slopes and rounded contours are deceiving, because geologically it is about as complex as any ridge can be. Many of you will know that following a good wet winter, the slopes of Gorman Ridge are covered by a gorgeous display of wildflowers, including golden California poppy and blue lupin. The bedrock in Gorman Ridge is as varied and mixed up as the flowers.

Inbound: Except for Hungry Valley beds on the immediate right, other items are behind you.

● Beyond the turnoff of Highway 138 (Palmdale) and Quail Lake Road, the freeway curves gradually into a northwesterly course, bringing more of Gorman Ridge into view. Gorman Ridge lies within the San Andreas fault zone, here several miles wide. However, the line of most recent displacement, 1857, lies within a trench or seam along the southwest base of the ridge. Clumps of green trees mark wet spots, usually closed depressions, and border sag ponds within this fault trench. These are features attributable to recent tectonic activity. You are proceeding to a junction with the San Andreas seam, and within two miles a long left curve in the freeway aligns your route directly along the San Andreas fault trough or seam. You reach this point beyond *"Gorman 3, Bakersfield 48, Sacramento 323"* at *"Gorman Exit 1 mile."* From there you are traveling on the trace of one of the largest and most active faults in the western hemisphere.

Keep watching Gorman Ridge. The occasional dark brown spots are outcroppings of volcanic rock, but there are many other kinds of rock within the geological smorgasbord composing the ridge. It's as though the San Andreas fault reached out and took a slice of just about everything that went past on the adjoining blocks and assembled them all in Gorman Ridge.

Inbound: Brown lava spots are easy to see in Gorman Ridge. Light areas are likely to be ground-up marble, sandstone, shale, or granitic rock.

● Within the seam, vegetated sag ponds lie just right of the freeway. Gorman is located directly on the most recent line of displacement, so hold your hat and hope nothing happens—unless you crave excitement.

Inbound: At Gorman and for 0.8 mile beyond, you travel right down the seam of the San Andreas fault zone, with the southwest face of Gorman Ridge in full view on the left. Sag ponds along the fault seam are made evident by clumps of trees.

● Frazier Mountain has been in view for some time and now lies left of you. With a summit elevation of 8,013 feet, it is one of the higher mountains of the area and is composed largely of old metamorphosed rocks. Partway up the flank facing you is a broad, gently sloping bench underlain by about 75 feet of brownish gray debris, seen well at 11:45 o'clock from the Gorman exit. This debris is alluvial gravel derived from Frazier Mountain, and it rests on light-colored beds of the Hungry Valley Formation. The old, dark metamorphic rocks rising above the bench on Frazier Mountain are thrust over the light-colored Hungry Valley beds. Frazier Mountain has been driven southeastward up and over the much younger Hungry Valley beds by movements that cannot be older than Pleistocene (1.8 m.y.). These thrust-fault relationships are exposed in some canyons on the south flank of Frazier Mountain that are not visible from the freeway, especially at 65 miles per hour.

Inbound: Frazier Mountain and the bench are well seen inbound from the Frazier Park turnoff.

● At *"Elevation 4,000 ft."* beyond Gorman, the freeway curves right on a course that carries you across and eventually out of the San Andreas fault zone. The most recent line of displacement is crossed obliquely just short of the Tejon Pass crest. It is marked by a vertical band of unusually dark material, about 20 feet wide, visible on both sides, but best on the left. All rocks in the highway cuts here are ground to a fine mush, the variegated colors representing different rock types. There's a lot of friction and grinding when this big fault moves, and the rocks are literally macerated, creating what geologists call *fault gouge.* The San Andreas would make a fine flour mill, only you would have to wait a while for its services; the last displacement here occurred during the Fort Tejon earthquake of 1857. This was a major shock, with an estimated magnitude in excess of eight on the Gutenberg-Richter scale.

Inbound: You enter the ground-up rock of the fault zone 0.1 mile beyond "Truck Brake Inspection Area Exit 3/4 Mile" and pass obliquely across it for 1.3 miles to "Elevation 4,000 ft." on the south side of Tejon Pass.

● About 100 yards beyond the dark gouge zone is the crest of Tejon Pass (4,144) and the start of a descent into Castac Valley. *Castac* is the officially approved spelling derived from a Native American word. Castaic may be a modern modification of Castac. To the left, approaching Frazier Park exit, the deep linear trough of the San Andreas lies between Frazier Mountain and Tecuya Ridge to the north. Look back left beyond the Frazier Park turnoff to get a better view of the bench on the east face of Frazier Mountain.

Inbound: These features are well seen approaching the turnoff to Frazier Park. People wishing to see San Andreas fault gouge close up can turn off at Frazier Park off-ramp and take the old four-lane highway to Gorman interchange.

● Shortly, beyond Frazier Park exit, the terrain opens to the northeast, and you see usually dry Castac Lake, about two miles to the right, and a long narrow valley beyond extending northeast and followed by a road. You are looking directly up another of southern California's great faults, the Garlock. In the opposite direction, the Garlock is headed west-southwestward on a collision course with the San Andreas. The San Andreas wins because the Garlock ends here after a 150 mile journey from the southern end of the Death Valley Trough, but it may have helped to put a considerable bend into the San Andreas in this area. As you can imagine, the intersection of two large faults causes much structural complexity. So much, that John P. Buwalda characterized this area as the "structural knot" of

California. The overthrust faults of Frazier Mountain are part of this complexity. The Garlock, like the San Andreas, is a lateral-slip fault, but the sense of slippage is opposite, left lateral, and smaller, possibly 40 miles compared to something between 150 to 300 miles on the San Andreas.

Inbound: The rest area is a good place from which to look east up the trace of the Garlock fault.

● Between "*Kern County Line*" and "*Lebec-Right Lane*" are good views at 1 o'clock over Castac Lake and up the trace of Garlock fault. This is also a stretch in which to look left again at Tecuya Ridge, and note that it, like Gorman Ridge, is made up of many rock types. However, they are rocks far different from those in either Frazier Mountain or Gorman Ridge, an indication of large lateral displacement on the San Andreas. Much of the light-colored rock in Tecuya Ridge is a white marble, which has been mined in small quarries farther west. The rest area on the right, 0.9 mile beyond the Frazier Park exit, makes a good observation point.

Inbound: The rest area provides a place to stop to inspect these features and relationships.

● Beyond the large curving overpass to Lebec, the freeway completes a swing left, back to its more normal northwesterly course, and you leave the San Andreas-Garlock complexities behind in favor of the relatively broad, simple, gentle floor of the northwest arm of Castac Valley. The hillslopes on both sides are made up of crystalline rocks, but they are so fractured and weathered, as shown by the road cut exposures, that good hillside outcrops are lacking until you get beyond "*Truck Brake Check Area 2 Miles Ahead on the Right.*" You remain in crystalline rocks, mostly old metamorphics, almost to the base of Grapevine Grade except for local masses of landslide breccias.

Exposures of fractured crystalline rocks have been seen in both walls of Grapevine Canyon well before you passed Fort Tejon.

● Having now passed safely across the San Andreas fault, you should realize that you are in a different world, geologically speaking. The area southwest of the San Andreas is on the Pacific plate, and you are now on the North American plate. The geology, some people would also say the life style, of the country southwest of the San Andreas, is somewhat exotic compared to that of much of the rest of North America. The Pacific and North American plates move past each other at a current average rate of about two inches per year, the Pacific plate moving relatively northward. Release of strain stored in rocks by this movement, like coiling a spring, causes many of California's major earthquakes.

Inbound: Plate tectonics is a powerful geological concept, but seeing concrete evidence of its workings is difficult. Geological structures and topographic features seen here have been created by plate movements.

● About 1.5 miles beyond the brake check area and beyond historical Tejon Ranch (1843) is an overpass, and 0.6 mile beyond you have the opportunity of taking an off-ramp and route leading across I-5 to the site of old Fort Tejon. This is now a state park, a good place for a picnic lunch, and an interesting stop for any aficionado of early western history. There's a nice little museum display recounting historical high points and a restoration of barracks and officers' quarters to be seen. Fort Tejon was established in 1854 and served as a base for the 1st U.S. Dragoons for about a decade. It housed many young officers who became famous generals in the Civil War. The U.S. Army Camel Corps also operated from here.

Inbound: Watch for roadside signs directing you to Fort Tejon. Access is easier than outbound.

● Back on I-5 northbound, you descend Grapevine Canyon (*Cajon de las Uvas*). The walls get progressively higher and impressively steeper until you start to break out into San Joaquin Valley. Rocks in the walls are largely varieties of old crystalline types, interrupted in places by accumulations of coarse angular rock-slide breccias. A good example is on the right 1.3 miles beyond the last overpass. In the growing season you can see bright green clumps of native grapevines on the lower canyon walls, on the right near *"Elevation 2,000 Ft."* and on the left downgrade from the lower watering stop. In places, the walls of Grapevine Canyon are made up of gigantic angular boulders. These are old rock-slide deposits.

Inbound: The lack of layering and the large size and angularity of the fragments in these breccias are the key to their landslide origin. The grapevine patches are easy to spot in the growing season. Outcrops of crystalline rock are good once you enter the steep narrow defile of Grapevine Canyon.

● Near the base of Grapevine Grade the canyon widens. A careful look at road cuts and hillsides at and beyond Call Box 88 shows sedimentary rather than crystalline rocks, mostly brownish sandstones and some brown-to-gray shales of Eocene (37 to 53 m.y.) and Miocene (5 to 23 m.y.) age. The sedimentary rocks begin about where the two branches of the freeway separates.

Inbound: Eocene sandstones and shale outcrop on the right upgrade beyond a slide breccia and continue to the joining of freeway lanes.

● Near the bottom of the grade, 0.3 mile beyond *"Grapevine 1 Mile"* at 9 o'clock between branches of the freeway is a nice 20 foot fault scarp with two water tanks perched on its brink (Photo T–6). Here, you cross the Pleito thrust fault, which determines the northern base of the mountains.

Inbound: Southbound travelers can see the fresh little scarplet on the left between freeway branches at 9 o'clock approaching and at "Lebeck 8, Castaic 43, Los Angeles 82."

Photo T–6.
Tanks are perched above young fault scarp at base of San Emigdio Mountains. Locale is between freeway lanes of Interstate 5 at south edge of Grapevine.

The first big road cut beyond, on the right, exposes coarse angular rock-slide breccia, with crystalline fragments, resting on Tertiary sandstone.

● Beyond the restaurants and service stations at Grapevine, you move onto the relatively flat floor of San Joaquin Valley. At 2.2 miles beyond Grapevine the California Aqueduct is crossed. About 6 miles right lies the gigantic Edmonston Pumping Station (not visible); the road to it turned off at Grapevine. This plant lifts aqueduct water fully 1,900 feet so it can cross the Garlock and San Andreas faults on the surface rather than in a tunnel deep underground. A 20 foot lateral displacement, as occurred in the 1857 Fort Tejon earthquake, would play havoc with an underground water tunnel, possibly shutting it off completely. Such considerations made the California Aqueduct project an engineering geologist's dream, or nightmare, according to your outlook.

Inbound: Initial plans called for just such a tunnel to avoid the pumping effort.

● In the vicinity of Laval Road *("Lamont-Lake Isabella exit"),* a mile north of the aqueduct, are a number of pumping oil wells just right of the freeway, one of the San Joaquin Valley's many oil fields.

Inbound: If anything, the oil wells are easier seen inbound on the left.

● A little before separation of I-5 and Highway 99, about 1.2 miles north of Laval Road, you pass the east end of Wheeler Ridge (see Photo 2–17). Although the guide for this segment ends at the separation, northbound travelers on either I-5 or 99 are urged to look back toward the north face of Wheeler Ridge.

Inbound: Southbound travelers on either freeway get an even better and easier view.

Wheeler Ridge is a spur off the Pleito Hills extending eastward almost to the freeway. It is an unusual geological feature in many ways, but we will focus on just a few of its aspects. The epicenter of the 1952 Tehachapi-Arvin earthquake lay just south of Wheeler Ridge, and the strain released during that shock—as well as a huge number of aftershocks—progressed northeasterly from Wheeler Ridge along the Whitewolf fault.

One feature likely to catch your eye on the north side of Wheeler Ridge is a large sand and gravel operation near its eastern end. There sand and gravel are being excavated from the Pleistocene (250,000 years) Tulare Formation, which is involved in the Wheeler Ridge folding. Wheeler Ridge is so young—less than 250,000 years—that its present form reflects faithfully the configuration of the anticlinal deformation that raised it. Most folds in the United States have suffered so much erosion, either during or following their creation, that the original topographic configuration created by deformation is erased or modified beyond easy recognition.

Inbound: The fact that beds of sand and gravel folded into an anticline are loose and unconsolidated enough to be excavated easily is most unusual. This area is accessible off the end of Sabodan Road.

The next thing likely to catch your eye are four large pipes rising from the Wind Gap (or I. J. Chrisman) pumping station on the California Aqueduct and passing over the top of Wheeler Ridge by means of a broad, open, U-shaped saddle, which geologists call a *wind gap.* Wind gaps can be created in different ways, but this one was formed as follows.

Inbound: There is no mistaking these pipes. For a closer view get onto Highway 166 and turn south on Sabodan Road toward Wind Gap pumping station.

Before there was any Wheeler Ridge, streams from the high San Emigdio Range, on the south skyline, flowed northward into San Joaquin Valley over the future site of Wheeler Ridge. As the ridge began to form slowly by anticlinal folding (bowing upward) some of the smaller streams were turned aside because their cutting power was exceeded by the rate of uplift.

Larger streams were able to maintain their course, and after a time they were flowing across the rising ridge in canyons. The stream that formerly occupied this wind gap finally gave up the struggle when Wheeler Ridge had attained about half its present height, either because the rate of uplift accelerated or the stream found the deeper rocks more resistant and their increasing breadth so great that it could not cut down as rapidly as the ridge rose. Or perhaps a spell of dry climate reduced its flow and power to cut down.

Whatever the cause, it was turned aside and ever since has flowed in the canyon of Salt Creek to the east, where the ridge is not so high. A gap of this last type, through which a stream still flows, is called a *water gap.* The old abandoned gap is, as you now surmise, a wind gap, because only wind flows through it. Continued uplift of Wheeler Ridge has raised this wind gap several hundred feet above the level it occupied when it was a water gap.

Wind gaps have played a significant role in United States history. Daniel Boone built his road to Kentucky from Virginia through Cumberland Gap, a large wind gap in the Appalachian Mountains. Robert E. Lee's Confederate army got up to Gettysburg, in part, by using both wind and water gaps.

Thanks to the activities of humans, the Wheeler Ridge wind gap is once again carrying water, but now in huge pipes and in the reverse direction. One wonders whether the engineers and administrators of the California Water Project ever pause and pay respects to the unnamed stream that gallantly struggled against the rising Wheeler Ridge and won the battle long enough to create the wind gap they now use for easier transport of water into southern California.

Inbound: Read the three preceding paragraphs. No special navigational points are needed. Sabodan Road is accessible from Highway 166, which in turn is accessible from either I-5 or Highway 99.

Appendix A

Geological Time Scale

Era	Period	Epoch	Starting Age
Cenozoic	Quaternary	Holocene	
			10,000 yrs.
		Pleistocene	
			1.8 m.y.
	Tertiary	Pliocene	
			5
		Miocene	
			23.5
		Oligocene	
			39
		Eocene	
			53.5
		Paleocene	
			65 m.y.
Mesozoic	Cretaceous		
			144
	Jurassic		
			208
	Triassic		
			245 m.y.
Paleozoic	Permian		
			286
	Pennsylvanian		
			320
	Mississippian		
			360
	Devonian		
			408
	Silurian		
			438
	Ordovician		
			505
	Cambrian		
			570 m.y.
Precambrian	Proterozoic		
			2500 m.y.
	Archean		
			3950 m.y.
-------Lost Interval-------			
Origin of Earth			4600 m.y.

Appendix B

Glossary

Aa. Lava with a blocky, jagged, clinkery, spinose surface.

Abrasion. The mechanical wearing of solid materials by impact and friction.

Acidic. Used here to describe igneous rock composed largely of light-colored minerals and more than 60 percent silica.

Agglomerate. A fragmental volcanic rock consisting of large, somewhat rounded stones in a finer matrix, much like conglomerate in appearance but wholly volcanic in constitution.

Alluvial. *See* Alluvium.

Alluvium. Unconsolidated gravel, sand, and finer rock debris deposited principally by running water; adjective is *alluvial*.

Angular unconformity. An arrangement in which older, deformed stratified rocks have been truncated by erosion and younger layers have been laid down upon them with a different angle of inclination.

Antecedent stream. A stream that maintained its course despite localized uplift across its path; the stream anteceded the structure.

Anticlinal core. The mass of older rock in the heart of an anticline.

Anticlinal nose. The place where beds at the axis of a plunging anticline pass beneath the ground surface.

Anticline. A fold in stratified rock convex upward. Beds on the flanks are inclined outward.

Apron (alluvial). A broad, gently sloping alluvial surface at the foot of a mountain range, formed by coalescing alluvial fans.

Arroyo. The wide, flat-floored channel of an intermittent stream in dry country.

Ash. *See* Volcanic ash.

Asphalt. A dark brown to black, viscous hydrocarbon usually formed by the loss of volatiles from petroleum.

Asymmetrical fold. A fold in which one limb is more steeply inclined than the other.

Attitude. The orientation of a plane surface, commonly bedding, with respect to compass bearing and to the horizontal.

Axis.   The central line of an elongated geological structure such as an anticline or syncline.

Badlands.   Extremely rough barren terrain with unusually steep slopes and sharp divides, riven by narrow, steep-walled gullies.

Barchan.   An isolated, crescent-shaped dune, convex upwind.

Barranca.   A vertical-walled gully cut by an intermittent stream in relatively unconsolidated material.

Basalt.   A fine-grained black lava relatively rich in calcium, iron, and magnesium. The extrusive equivalent (in composition) of gabbro.

Basement.   Old crystalline rocks on which younger rocks have been deposited.

Basic.   Used here for darker igneous rocks relatively rich in iron, magnesium, and calcium, and containing about 50 percent or less silica.

Bastnaesite.   A greasy yellowish to reddish-brown carbonate (CO_3) mineral containing rare earth elements, particularly cesium, lanthanum, and other rare elements.

Batholith.   A large igneous body intruded into the earth's crust at considerable depth, where it cooled slowly to form coarsely crystalline rock.

Beach.   A narrow, constantly shifting deposit of sand or stones between high and low water levels along the shore of a standing water body.

Bedding.   The layered structure of sedimentary rocks.

Bedding plane.   A planar surface separating successive depositional layers within sedimentary rocks.

Bedrock.   Consolidated rock material of any sort.

Bench.   A level or gently sloping area interrupting an otherwise steep slope.

Benchmark.   An established mark, the elevation of which is accurately determined with respect to sea level.

Berm.   A narrow, linear embankment, or a bench, shelf, or ledge that breaks a slope, natural or artificial.

Bluff.   A bold, broad, high rock face or cliff, overlooking a flat area.

Brea.   Viscous asphalt deposit formed around an oil seep owing to evaporation of volatiles.

Breakaway-scarp.   Steep face at the head of a landslide, slump, or earthflow, left when the moving mass pulled away.

Breccia.   A rock containing abundant angular fragments of rocks or minerals. Types include sedimentary, volcanic, tectonic, landslide, and others.

Brink.   Upper edge of a steep declivity, not necessarily the highest part of the terrain.

Calcareous.   Rich in calcite.

Calcite.   A common mineral composed of calcium, carbon, and oxygen ($CaCO_3$). The principal source of cement.

Caliche.   A calcareous deposit formed within dry-region soils by weathering.

Capture.   *See* stream capture.

Carbonate rocks.   Rocks composed of the minerals calcite (calcium carbonate) and dolomite (calcium-magnesium carbonate).

Carbonatite.   A carbonate rock seemingly of igneous (controversial) origin.

Cavernous weathering.   Action of a natural force on a rock face that forms hollows, niches, small caves, or other recesses.

Chamise.   A woody shrub, 2 to 8 feet high, with distinctive bundles of linear needle-like leaves, somewhat resinous; a common constituent of many chaparrals.

Chaparral. A dense complex of scrubby bushes, including a variety of plants, covering hillsides in semiarid environments that have 12 to 25 inches of seasonal precipitation.

Cinders. See Volcanic cinders.

Clast. Rock or mineral fragment within course sedimentary rocks.

Clastic. Rocks or minerals broken into fragments.

Clastic dike. A tabular body of fragmental rock material transecting country rock, commonly sedimentary (for example, sandstone dikes intruding shale).

Clay. A group of minerals with a strongly layered structure usually formed by alteration of aluminum-rich parent minerals. Also used for mineral fragments less than 1/256 mm diameter.

Cleavage. The facility to break along parallel smooth planes, especially in minerals but also in rocks.

Climbing dune. A dune-size accumulation of wind-blown sand driven upslope before the wind.

Closed depression. Depression on the ground surface with a rim everywhere higher than the floor.

Columnar jointing. The joint pattern that separates rock into polygonal columns.

Conchoidal fracture. The fracture habit of rocks or minerals producing a smooth, curved, commonly concave surface.

Concretion. A nodular mass within a sedimentary rock distinguished by greater coherence, density, color, mineralogy, or other properties.

Conglomerate. A sedimentary rock consisting of larger, rounded rock and mineral fragments embedded in a finer, usually sandy, matrix and all cemented together.

Core stone. The roughly spheroidal core of sound rock within a partly disintegrated subsurface joint block of a massive parent rock.

Country rock. The prevailing bedrock of a region into which younger rock or mineralization has been introduced.

Crop out. An exposure (outcropping) of bedrock on the surface.

Cross bedding (or lamination). An internal bedding within a sedimentary layer that forms a distinct angle with the upper surface of that layer or with other, differently oriented cross beds within it.

Crystal. A regular, solid, geometrical form bounded by plane surfaces expressing an internal ordered arrangement of atoms.

Crystalline rocks. A term commonly applied to mixed igneous and metamorphic rocks, or to either separately.

Cuesta. An asymmetrical ridge with a steep face on one side and a gentle smooth slope on the other, formed by erosion of gently tilted sedimentary beds of different resistance.

Debris. Broken-up and usually partly decomposed rock materials.

Debris cone. A cone-shaped accumulation of rock debris at the mouth of a gully or small canyon, usually smaller, steeper, and often rougher than an alluvial fan.

Debris flow. A flow of usually wet, muddy rock debris of mixed sizes, much like a slurry of freshly mixed concrete pouring down a chute.

Decomposition. The chemical breakdown of rocks and minerals.

Desert pavement. An armor of closely fitted stones, one layer thick, on the surface of alluvial material. Basically, a residual accumulation of larger fragments because of the removal of fine particles.

Desert varnish. A thin coating rich in iron and manganese on rock surfaces developed by weathering.

Diatomite. A sedimentary rock consisting almost entirely of the siliceous skeletons of single-celled algae.

Dike. A sheet-like body of igneous rock formed by intrusion along a fracture.

Diorite. A coarse-grained intrusive igneous rock about midway between a granite and a gabbro in chemical and mineralogical composition.

Dip. The direction and degree of inclination (from horizontal) of a sedimentary bed or any other geological planar feature.

Dip slope. A smooth, inclined surface formed by the exposed bedding plane of a sedimentary layer; the back-slope of a cuesta.

Disconformity. An unconformity within a sedimentary sequence with the beds above and below the erosion surface parallel.

Disintegration. The physical breakup of rocks and minerals.

Distributaries. Diverging channels in a stream by means of which the stream distributes water and debris, as across an alluvial fan or delta.

Dolomite. A sedimentary rock composed of the mineral dolomite, a calcium-magnesium carbonate.

Dome. (Topographical) A roughly circular, upwardly convex land form. (Structural) In sedimentary rocks, an outward dip or inclination of the beds in all directions. (Volcanic) A circular, convex extrusion of highly viscous lava.

Dune. A deposit of wind-blown sand of a size and shape capable of capturing additional sand.

Earth flow. A form of mass movement in which relatively unconsolidated surface material, usually weathered, flows down a hillside.

Embayment. An indentation along a shoreline, mountain front, or any other natural linear feature.

End moraine. A moraine deposited at the lower end of an ice stream or outer end of an ice lobe.

Epicenter. The spot on the earth's surface directly above the subsurface point at which an earthquake shock originates.

Erosion. The removal of rock material by any natural process.

Extrusive rock. The rock brought up onto the earth's surface, usually in molten condition (lava).

Fan. A deposit, usually alluvial, of rock debris at the foot of a steep slope (mountain face) with an apex at the mountain base (canyon mouth) and a radial, fan-like, divergence therefrom.

Fanglomerate. The consolidated deposits of an alluvial fan; a variety of conglomerate that is coarse and ill-sorted, and that contains angular stones.

Fault. A fracture along which blocks of the earth's crust have slipped past each other.

Fault plane. The two-dimensional surface along which fault displacement has occurred.

Fault ridge. An elevated, elongate block lying between two essentially parallel faults.

Fault slice. A narrow segment of rock caught between two essentially parallel, closely adjacent faults.

Fault zone. A zone in the earth's crust consisting of many roughly parallel, overlapping, closely spaced faults and fractures; may be up to several miles wide.

Feldspar. An abundant rock-forming class of minerals composed of aluminum, silicon, oxygen, and one or more of the alkalies, sodium, calcium, and potassium.

Femic.   Minerals or rocks of dark color especially rich in iron, magnesium, and calcium.

Ferromagnesian.   Dark rocks and minerals rich in iron and magnesium (mafic).

Flatiron ridge.   A linear ridge with one very smooth flank formed by erosion of tilted sedimentary rocks and given a triangular shape by cross-cutting canyons.

Flood plain.   A strip of low, smooth land bordering a river underlain by silts deposited during overbank floods.

Flute.   A small, elongate, scoop-shaped erosional depression, common on surfaces of wind blasted stones.

Fluvial.   Features of erosion or deposition created by running water.

Foliation.   A crude banding formed in rocks by metamorphism, less regular than the bedding of sedimentary rocks.

Foot wall.   The underside of an inclined fault.

Foredune.   A linear sand-dune ridge formed along the edge of a lake or the ocean, usually inland from a beach.

Formation.   A geological formation is a rock unit of distinctive characteristics that formed over a limited span of time and under some uniformity of conditions. To a geologist it is a rock body of some considerable areal extent that can be recognized, named, and mapped.

Gabbro.   A dark, coarse-grained intrusive igneous rock richer in iron, magnesium, and calcium and poorer in silica than granite.

Geophysical exploration.   The subsurface exploration of rocks and structures carried on by indirect means such as gravity or magnetic variations.

Geophysics.   The discipline that treats the physics of the Earth.

Geothermal.   Involving heat from within the earth.

Gneiss.   A coarse-grained metamorphic rock with irregular banding (foliation).

Gorge.   A narrow, steep-walled passage cut into rock by a stream.

Gouge.   Finely ground rock material within a fault zone.

Graben.   A sizable block of the earth's crust dropped down between two faults steeply inclined inward, giving a keystone shape to the block, longer than it is wide.

Grain.   A term used here for a perceptible linear pattern in landscape features of a region, usually reflecting a similar pattern in underlying rock structure.

Granite.   A common, coarse-grained, igneous intrusive rock relatively rich in silica, potassium, and sodium.

Granitic.   One of many coarse-grained igneous intrusive rocks not strictly of granite composition.

Granodiorite.   A coarse-grained, igneous intrusive rock half way between a granite and a diorite on the scale of rock composition.

Gravel.   The natural, unconsolidated accumulation of mixed sand and stones, in which constituents are usually somewhat rounded.

Gravity fault.   An inclined fault on which the overhanging block has moved relatively downward.

Groove.   The narrow linear depression of relatively uniform width and depth commonly formed by erosion on rocks or sediments. Small ones are created by sand blasting by wind.

Ground water.   Water filling pores and other openings in subsurface rock materials within the zone of saturation.

Gully.   A small ravine cut by running water.

Gypsum.   A common mineral, hydrated calcium sulfate, used to make plaster.

Hanging valley. A tributary valley the floor of which is much higher at its mouth than the floor of the trunk valley.

Hogback. A ridge composed of a resistant layer within steeply tilted eroded strata.

Ice Age. The interval in Earth history when glaciers of continental proportions formed on land.

Igneous rocks. A class of rocks formed by crystallization from a molten state.

Inclusion. A fragment of older rock inclosed (included) within an igneous rock.

Incompetent. A rock that is relatively weak and that responds readily to pressure by crumpling or by flow.

Intermittent stream. A stream that does not have a continuous or perennial flow.

Intrusive. Rocks or rock masses that have been intruded or injected into other rock, usually in a molten state.

Joint. The plane surface of a fracture, without displacement, within a rock.

Jointing. A family of joints within a rock mass.

Kipuka. Hawaiian term for an island of any sort completely surrounded by a sea of younger lava.

Landform. Any topographic feature of the land surface, of natural origin with distinctive and consistent characteristics.

Landslide. A phenomenon involving rapid, often catastrophic downslope movement by gravity of rock or rock debris primarily by basal slippage.

Lateral fault. A fault on which the displacement is sidewise rather than up-down.

Lateral moraine. A ridge-like deposit of bouldery ill-sorted debris laid down along the lateral margin of a valley glacier.

Lava. Either the molten rock material extruded onto the earth's surface and or the consolidated (crystallized) rock from that molten material.

Left-lateral fault. A fault on which the opposing block appears to have moved to the left, no matter on which side of the fault you stand.

Limb. One of the two sides of an anticline or syncline.

Limestone. A sedimentary rock composed wholly or almost wholly of the mineral calcite.

Lithology. The characteristics of a rock's class, composition, color, texture, and structure.

Macadam. A mixture of sand and fine gravel bonded by asphalt.

Magma. The molten rock within the earth's crust.

Marble. Recrystallized limestone or dolomite; a metamorphic rock.

Marine. The ocean environment; marine sediments are those deposited in the ocean.

Marine terrace. A landform consisting of a flat tread with a cliff on its inner edge bordering a sea shore, created by marine erosion.

Mass movements. The movement, usually downslope, of a mass of rock or rock debris by gravity, not transported by some other agent such as ice or water.

Matrix. The fine-grained constituents of a rock in which coarser particles are embedded.

Mesa. A flat-topped tableland with steep sides.

Mesozoic. One of the eras of the geological time scale (see Appendix A) extending from 65 to 245 m.y. ago.

Metamorphic rocks. Rocks that have undergone such marked physical change because of heat or pressure or both as to be distinct from the original rock. The process is *metamorphism*.

Metavolcanic. Rocks formed by metamorphism of volcanic materials.

m.y. An abbreviation for a million years.

Mineral. A homogeneous, naturally occurring, solid substance of inorganic composition, consistent physical properties, and specified chemical composition.

Mineralized rock. Rock impregnated with introduced mineral matter.

Mio-Pliocene. A time period for phenomena that spread-eagle the time interval between Miocene and early Pliocene.

Monolithologic breccia. A rock containing and formed of fragments of only one kind of rock.

Monominerallic rock. A rock composed of only one mineral; for example, limestone and dolomite.

Moraine. A deposit of coarse, ill-sorted rock debris laid down by glacial ice without intervention of other agents.

Mud pot. A shallow, hot-spring pit filled with bubbling mud.

Mudflow. A form of mass movement involving the flow of mud, usually containing coarser rock debris, in which instance the term debris flow is equally applicable.

Mudstone. A fine-grained sedimentary rock that is hard to characterize as shale or siltstone because of massiveness or poor sorting.

Nodule. A small, hard, lump of mineral matter enclosed within a sedimentary matrix; also, a mineral knot in metamorphic rocks.

Normal fault. An inclined fault on which the overhanging block has moved relatively down; a gravity fault.

Nose. *See* Anticlinal nose.

Oblique air photo. A photo taken with the axis of the camera tilted from vertical. If the horizon shows, it is called a *high-oblique photo.*

Obsidian. Natural volcanic glass; lava that cooled so rapidly that it didn't crystallize.

Odometer. An instrument for measuring distance.

Ore deposit. An accumulation of metallic minerals that can be mined at a profit. The minerals are termed *ore minerals,* and the aggregate is termed *ore.*

Outcrop. An exposure of bedrock at the surface.

Outlier. A detached, free-standing mass of rock separated by erosion from the principal body of the same rock.

Overturned. Strata in which the age sequence is reversed, old above young.

Paleomagnetism. The natural remnant magnetism in rocks and minerals preserving a record of previous magnetic field orientation and polarity.

Paleozoic. A major era of the geological time scale embracing the interval from 245 to 570 m.y. (*see* Appendix A).

Pediment. A relatively smooth, gently sloping surface produced by erosion at the foot of a steeper face, usually a mountain.

Pegmatite. A very coarse-grained igneous rock formed by the fluids given off in the late stage of crystallization of an igneous body; most often close to granite in composition.

Pendant. A large mass of metamorphic rock within a younger intrusive rock, thought to have hung down into the original intrusive body from the roof of the intrusive chamber.

Placer deposit. A water-laid accumulation of rock debris containing a concentration of heavy, physically and chemically resistant, valuable mineral such as diamond, gold, or platinum. Such minerals are described as *placer minerals.*

Plain. Any flat, smooth area of at least modest extent at a low level.

Plate tectonics. The behavior and interaction of huge crustal plates on Earth as moved by forces acting in Earth's mantle.

Playa. The flat, smooth floor of a dry lake in desert regions.

Pleistocene.   An epoch within the Cenozoic Era of the geological time scale (*see* Appendix A); usually taken to embrace the last 1.8 million years.

Plio-Pleistocene.   Phenomena that bracket the time interval between late Pliocene and early Pleistocene.

Plug.   A small, cylindrical, near-surface, igneous intrusive body.

Plunge.   The inclination from horizontal of the long axis of a fold or warp.

Plutonic.   Igneous rocks or bodies formed at great depth.

Pluvial Period.   An interval of cooler, wetter conditions in a dry region essentially, coincident with a phase of glaciation in colder, better-watered areas.

Potassium-Argon.   A method of absolute dating of rocks and minerals using the ratio of radioactive potassium to its daughter product, the argon 40 isotope.

Pothole.   A narrow cylindrical hole worn into solid rock by a fixed vortex in a stream.

Precambrian.   All rocks older than Paleozoic (*see* Appendix A).

Pumice.   Frothy rock glass, so light that it floats.

Pyroclastic.   Hot or fiery (*pyro*) fragmental (*clastic*) debris thrown out of an explosive volcanic vent.

Pyroxene.   A common igneous- and metamorphic-rock family of minerals, often green to black, and ranging widely in composition.

Quartz.   One of the most common minerals, hard and chemically resistant, composed of silicon and oxygen (SiO_2).

Quartzite.   A rock formed by metamorphism of sandstone, which is hard and coherent, and consists of quartz.

Radioactive.   The property of some elements to spontaneously change into other elements with the emission of charged particles, usually accompanied by generation of heat.

Radio-carbon.   The radioactive isotope of carbon (14_C) that disintegrates at a known rate. It is used to determine geological ages up to about 40,000 years.

Rare earths.   The oxide compounds of rare-earth elements, such as cerium, ytterbium, neodymium, and others.

Recharge well.   A well that has been designed for injection of fluids into the ground.

Relief.   (Topographical) Difference in elevation of contiguous parts of a landscape, valley to peak.

Reverse fault.   An inclined fault on which the overhanging block has moved relatively upward.

Rhyolite.   An extrusive igneous rock of granitic composition, fine-grained, often light-colored to red.

Rift.   As used here, the shallow topographic trench, a mile or two wide, along the trace of a major fault.

Right-lateral fault.   A fault on which the displacement of the opposing block appears to have been to the right, no matter on which side of the fault you stand.

Rill.   A small, shallow, narrow, straight, unbranched channel extending directly downslope, eroded by an ephemeral streamlet.

Rilled.   A slope dissected by a family of parallel, evenly spaced rills.

Rillensteine.   A stone with small, interlacing, worm-like solution channels on its surface. They form on soluble rocks, most commonly limestone.

Riparian.   Pertaining to the banks of a water body, usually a stream. Commonly applied to stream-side vegetation.

Rock.   An aggregate of minerals.

Rock cleavage.   The facility to break along parallel, closely spaced, smooth planes within a mass of rock.

Rock glacier.   An accumulation of large angular blocks of rock, usually lobate in form with steep margins, that moves slowly by creep.

Rockfall. The relatively free fall of rock masses from steep bedrock faces.

Rock varnish. *See* Desert varnish.

Roundstone. Any rounded rock fragment larger than a sand grain. A common constituent of conglomerates.

Sag pond. A small water body occupying a closed depression created by recent activity along a fault zone.

Sandstone. A sedimentary rock formed by cementation of sand-size particles.

Scarp. A straight steep bank or face that can be a few feet to thousands of feet high, like the east face of the Sierra Nevada.

Schist. A finer-grained and more thinly and regularly foliated metamorphic rock than gneiss.

Scoria. Small fragments of porous volcanic rock thrown out of an explosive volcanic vent. Usually black or red and up to 1.5 inches in diameter.

Scour channel. A large, elongate, groove-shaped depression created by strong-current erosion in sedimentary deposits, usually filled with younger sediment.

Sea-floor spreading. The process wherein fluid magma wells up in the center of midoceanic ridges, creating new oceanic crust and shoving older crust outward on both sides.

Sedimentary rocks. A class of rocks of secondary origin, made up of transported and deposited rock and mineral particles and of chemical substances derived from weathering.

Sedimentary structures. The geometrical relationships created within sedimentary beds as they are laid down (example, ripple marks).

Seismology. The study of earthquakes.

Septum. An older mass of metamorphic rock separating two adjacent, intrusive, igneous bodies.

Serpentinite. A rock consisting largely of the mineral serpentine, a hydrous magnesium silicate, produced by alteration of igneous rocks rich in iron and magnesium.

Shale. A sedimentary rock consisting largely of very fine mineral particles, laid down in thin layers.

Silica. The hard, chemically resistant compound of SiO_2 (for example, quartz).

Siliceous. Rich in silica, SiO_2

Silicic. Rocks rich in silica.

Silicified. Material that has been largely replaced or impregnated by silica.

Sill. Tabular igneous intrusion conformable with the planar structure (bedding) of the host rock.

Siltstone. A fine-grained, well-bedded sedimentary rock composed of silt that is finer than sand and coarser than clay.

Slate. A weakly metamorphosed rock derived from shale by compaction with the development of closely spaced, smooth, parallel breaking surfaces (slaty cleavage).

Slickensides. Smooth, polished, striated surfaces produced by movement along a fault.

Slope wash. The soil and loose rock waste transported downslope by surface runoff; often mantles slopes.

Slump. A form of mass movement in which a coherent block of rock slips downslope on a concave slip surface, experiencing some backward rotation in the process.

Soapstone. A massive, soft, slippery rock composed of the mineral talc, a hydrous magnesium silicate.

Soil. A layer of weathered and altered rock debris mantling underlying parent material.

Sorting. The arrangement of particles by size.

Spur. The subordinate ridges extending from the crest of a larger ridge.

Strata. Layers of a sedimentary rock. Bedded rocks are *stratified*.

Stratigraphic. Relationships between successions of layers (strata) within a sequence of sedimentary beds.

Stratigraphy. A science involving the definition and description of natural subdivisions, principally within sedimentary rock sequences.

Stream capture. The diversion of the headwaters of a stream because of headward growth of an adjacent stream.

Stripping. The process of removing an overlying deposit to expose an underlying surface, frequently a bedding plane.

Structural trough. A topographic trough created by faulting or down warping.

Structure. Phenomena that determine the geometrical relationships of rock units, such as folds, faults, and fractures.

Submarine canyon. A canyon carved in the floor of the ocean, presumably by some submarine process.

Superimposed stream. A stream that has cut down through an overlying mantle into rocks of different character and structure.

Swale. A slight depression in the midst of generally level land.

Syenite. An intrusive, igneous rock much like granite but lacking or containing little quartz.

Syncline. A downfold in layered rocks that is concave upward. Beds on the flanks are inclined inward.

Talus. The accumulation of generally angular rock fragments, from small to large, lying at the base of a cliff or steep rocky slope from which they were derived.

Tectonic. Relating to the deformation of Earth's crust by forces originating within the earth.

Terrace. A geometrical form consisting of a flat tread and a steep riser or cliff. Stream terraces, lake terraces, marine terraces, and structural terraces are distinguished in geology.

Terrain. An area of Earth's surface with common characteristics.

Terrane. An extensive area of related rock outcropping.

Terrestrial. Deposits laid down on land as contrasted to the sea; terrestrial conditions as compared to marine conditions.

Tertiary. A period of the Cenozoic Era (*see* Appendix A) embracing the time from 65 to 1.8 m.y. ago.

Thrust fault. A gently inclined fault along which one block is thrust over another.

Thrust plate. The upper block of a thrust fault.

Till. An ill-sorted, mixed fine and coarse rock debris deposited directly from glacial ice.

Travertine. An accumulation of calcium carbonate formed by deposition from ground or surface waters, commonly porous and cellular.

Tuff. Compacted pyroclastic debris, largely ash. If rock fragments are numerous, it is a tuff-breccia.

Turbulence. A mode of fluid flow in which flow lines are oriented in all directions.

Unconformity. A surface of erosion separating younger strata from older rocks.

Varnish. *See* desert varnish.

Vein. A sheet-like deposit of mineral matter along a fracture.

Ventifact. A stone whose shape and surface characteristics have been modified by natural sandblasting.

Vertical air photo. A photo taken with the axis of the camera pointed straight down toward the ground.

Volcanic ash. The fine-grained (less than 1/8 inch diameter) volcanic debris, often glassy, explosively erupted from a volcanic vent.

Volcanic cinders. Cinders that are like volcanic ash but coarser, 1/8 to 1 inch. Their fragments are highly porous.

Volcanic tuff. A compacted deposit consisting of ash, cinders, and occasionally larger fragments of solid volcanic rock. If the latter are numerous, it is known as a *tuff-breccia*.

Volcanism. A process by which magma and gas rise to the surface and are erupted.

Volcano-clastic. Volcanic rock that has been broken to pieces by violent eruption.

Warp. A part of the earth's crust that has been broadly bent.

Water gap. A gap in a ridge still occupied by the stream that cut it.

Water table. The level beneath the ground surface below which all openings in rocks are filled with water.

Weathering. The mechanical breakup and chemical decomposition of rocks on or near Earth's surface through interactions with the atmosphere, hydrosphere, and biosphere.

Wind gap. A gap or saddle in a ridge now abandoned by the stream that cut it.

Wineglass canyon. A canyon cut into the steep face of a mountain range. The fan at the mountain foot is the base, the gorge approaching the canyon mouth is the stem, and the headwaters basin is the bowl.

Woolsack weathering. A natural process that produces large rounded surface boulders in areas of homogeneous, coarse grained, granitic rocks that are jointed.

Zircon. A mineral found in small amounts in many igneous and metamorphic rocks, a zirconium silicate and a gemstone. Chemically and mechanically tough.

Appendix C

Annotated Bibliography

GENERAL BACKGROUND

Birkeland, Peter W., and Larson, Edwin E. *Putnam's Geology.* New York and Oxford: Oxford University Press, 1989.
A well-written, classical text on physical geology, nicely revised and updated by Birkeland and Larson.

Court, Arthur, and Campbell, Ian. *Minerals: Nature's Fabulous Jewels.* New York: Harry N. Abrams, Inc., 1974.
A book of spectacular photographs of beautiful minerals accompanied by short, pithy, informative paragraphs on each species. It's a joy just to leaf through the pages.

Easterbrook, Donald J. *Surface Processes and Landforms.* New York: Macmillan Publishing Co., 1993.
A current, well-illustrated and comprehensive text on the many processes shaping landforms on the face of Earth. An excellent presentation that is easy to understand.

Fenton, C. A., and Fenton, M. A. *The Rock Book.* New York: Doubleday, Doran and Co., 1940.
Although old, this highly readable book provides sound information on rocks in a style easily digested by nonprofessionals.

Shelton, John S. *Geology Illustrated.* San Francisco: W. H. Freeman and Co., 1966.
A superb collection of mostly low-elevation oblique air photos of a wide variety of geological features, many in southern California, with an accompanying explanatory text.

Tennissen, Anthony C. *Nature of Earth Materials.* Englewood Cliffs, N.J.: 1983.
This is a comprehensive, not too complex or professional treatment of rocks and minerals. It's an easy-to-use tool.

SOUTHERN CALIFORNIA (GENERAL)

Bailey, Edgar H., ed. *Geology of Northern California,* Bulletin 190. California Division of Mines and Geology, 1966.
A companion volume to the southern California book. Chapters 5 (Great Valley), 6 (Coast Ranges), and especially 4 (Sierra Nevada) are of particular interest in relation to areas described in this book.

Downs, Theodore. *Fossil Vertebrates of Southern California.* Berkeley: University of California Press, 1968.
A small pocketbook in popular style for nonprofessional readers; contains some striking color illustrations (paperback).

Geologic Map of California (Large scale). California Division of Mines and Geology.
This map consists of 26 separate sheets on a scale of 4 miles to the inch. The sheets can be separately purchased from the California Division of Mines and Geology, P.O. Box 2980, Sacramento, CA 95812–2980. The sheets of primary use in the areas treated herein are as follows: Mariposa, Fresno, Death Valley, Bakersfield, Trona, Kingman, Los Angeles, San Bernardino, Needles, Santa Ana, and Salton Sea. This is the finest state geological map extant.

Geologic Map of California (Small scale). U.S. Geological Survey and California Division of Mines and Geology, 1966, $3 prepaid.
This excellent, small-scale map showing natural provinces, faults, and rock distribution can be obtained by an order addressed to the same office as the large-scale map.

Hinds, N. E. A. *Evolution of California Landscape,* Bulletin 158. *California Division of Mines and Geology,* 1952.
Describes features within the various natural provinces; lavishly illustrated with photographs.

Iacopi, Robert. *Earthquake Country.* Menlo Park, Calif.: Lane Book Co., 1964.
A reliable description of California faults and earthquakes prepared for the layperson, with a good format and excellent illustrations.

Jahns, R. H., ed. *Geology of Southern California,* Bulletin 170. California Division of Mines and Geology, 1954.
This is the definitive professional work on southern California's geologic features. It is a massive compilation of articles, maps, and guidebooks dealing with almost every conceivable aspect of the region's geology.

Norris, Robert M., and Webb, Robert W. *Geology of California (Second edition).* New York: John Wiley and Sons, Inc., 1990.
This is the best book ever published on the regional geology of California. It has been recently and thoroughly revised and updated by Norris. The treatment is organized around natural provinces and is thorough and competent.

Russ Leadabrand Guidebooks. Los Angeles: Ward Richie Press.
An inexpensive series of small pocketbooks describing aspects of the natural and human history of various parts of southern California by a man who knows the country like the palm of his hand and loves it deeply. The following are particularly pertinent to areas or features treated herein:
A Guidebook to the San Gabriel Mountains, 1964.

A Guidebook to the San Bernardino Mountains, 1964.

A Guidebook to the Mountains of San Diego and Orange Counties, 1972

A Guidebook to the Mojave Desert (including Death Valley), 1966

A Guidebook to the Southern Sierra Nevada, 1968

Exploring California Byways:

I: Kings Canyon to Mexican Border, 1967

III: Desert Country, 1969.

SPECIFIC AREAS

Transverse Ranges

Fife, B. L., and Minch, J. A. *Geology and Mineral Wealth of the California Transverse Ranges.* Santa Ana, Calif.: South Coast Geological Society, 1982.
A compendium of many articles by many authors.

Peninsular Ranges

Peterson, G. L., et al. *Geology of the Peninsular Ranges.* California Division of Mines and Geology, Mineral Information Service (now California Geology), v. 23, pp. 124–127, 1970.

Los Angeles Basin

Woodring, W. P., Bramlette, M. N., and Kew, W. S. W. *Geology and Paleontology of Palos Verdes Hills,* Professional Paper 207. U.S. Geological Survey, 1946.
The recognized standard reference on Palos Verdes Hills by three very competent men.

Yerkes, R. F., McCulloh, T. H., Schoellhamer, J. E., and Vedder, J. G. *Geology of the Los Angeles Basin, California,* Professional Paper 420–A. U.S. Geological Survey, 1965.
The most authoritative introduction to the geology of this area ever prepared.

Mojave Desert

Dibblee, Thomas W., and Hewett, D. Foster. *Geology of the Mojave Desert.* California Division of Mines and Geology, Mineral Information Service (now California Geology), v. 23, pp. 180–185, 1970.

Death Valley

Hunt, Charles B., and Mabey, Don R. *Stratigraphy and Structure, Death Valley, California,* Professional Paper 494–A. U.S. Geological Survey, 1966.
The most modern, thorough, and competent professional geological publication available on the valley and its immediate environs.

Hunt, Charles B. *Death Valley: Geology, Ecology, Archeology.* Berkeley, Calif.: University of California Press, 1975.

Southern Coast Ranges

Page, Ben. "The Southern Coast Ranges." In *Geotectonic Development of California* (Rubey Volume). Englewood Cliffs, N.J.: Prentice Hall, pp. 329–417, 1981.

The Great Valley

Hackel, Otto. "Summary of the Geology of the Great Valley." In *California Division of Mines and Geology,* Bulletin 190, pp. 217–238, 1960.
A digest of both the Sacramento and San Joaquin valleys. A comprehensive overlook.

Owens Valley— Sierra Nevada

Bateman, P.C., and Wahrhaftig, Clyde. "Sierra Nevada Province." Chapter IV in *Geology of Northern California,* Bulletin 190. California Division of Mines and Geology, 1966.
This is the most authoritative statement available concerning the geology of the Sierra Nevada.

Hill, David W. *Inyo Skyline,* Bishop, Calif.: Chalfant Press, 1965.
A road-log guide to topographic and other features seen on a highway trip through Owens Valley (paperback).

————. *Mono Skyline.* Bishop, Calif.: Chalfant Press, 1969.
A road-log guide to topographic and other features north from Bishop (paperback).

Rinehart, C. D., and Ross, D. C. *Geology and Mineral Deposits of the Mount Morrison Quadrangle,* Sierra Nevada, California. Professional Paper 385. U.S. Geological Survey, 1964.
This publication is cited as an example of the many excellent professional papers and bulletins issued by the U.S. Geological Survey. The area is adjacent to the U.S. Highway 395 route to Mammoth.

Schumacher, Genny, ed. *Deepest Valley: A Guide to Owens Valley, Its Lakes, Roadsides, and Trails.* Berkeley: Wilderness Press, 1978.
An authoritative, well-written, and nicely illustrated guide to the natural and human history of Owens Valley and environs (paperback).

————, ed. *The Mammoth Lakes Sierra: A Handbook for Roadside and Trail.* Berkeley: Wilderness Press, 1969.
Equal in caliber to the booklet on Owens Valley and fulfilling a similar function for the Mammoth Lakes region (paperback).

Salton Trough

Dibblee, Thomas W. "Geology of the Imperial Valley Region." In *California Division of Mines and Geology, Bulletin* 170, pp. 21–28, 1954.

Biehler, Shawn, and Rex, Robert W. "Structural Geology and Tectonics of the Salton Trough, Southern California." In *University of California Riverside Museum Contribution No. 1*, pp. 30–42, 1971.

Basin Ranges

Anonymous. *Geology of the Basin Ranges.* California Division of Mines and Geology, Mineral Information Service (now California Geology), v. 21, pp. 131–133, 1968.

Index

Index of Localities and Features

Garlock road, 126
Garnet Hill, 234, 237
Gavin Canyon, 242
George C. Page Museum, 31
George F. Canyon (Palos Verde
 Hills), 34
Gerkin Road, 188–89
Gertsley borax mine, 99
Geysers, 50, 204
Ghost Town Road, 84, 86
Gibraltar dam and reservoir, 20
Gilman Springs Road, 229–30
Glacier Lodge, 186
Glamis, 46
Glass Mountain, 201
Golden Canyon, 118
Golden Queen mine, 147
Golden State Freeway, 57–58,
 137, 257
 Bucket of Worms interchange
  of Highway 14 and,
  137, 242
 fault scarp between lanes
  of, 257
 segment T from San Fernando
  Pass to San Joaquin Valley
  along, 242–58
 Transverse Ranges divided
  by, 20
Golden Trout Creek, 62
Goodsprings (Mr. Good's
 springs), 215
Gorman (village), 245, 249, 254
Gorman Creek, 253
Gorman Ridge, 253–54
Gower Gulch, 118
Grand Canyon, 217
Granite Mountains, 91, 207
Grapevine (settlement), 257
Grapevine Canyon (Sierras), 161
Grapevine Canyon (Tehachapi
 Mountains), 255–56
Grapevine Grade, 57–58, 255–56
Grapevine Mountains, 65
Great Basin region, 64–65
Great Valley, 56–58
Greenwater Range, 99
Greenwater Valley, 98–99
Griffith Park, 31
Gulf of California, 44, 46

H

Haiwee dam, 167–68
Haiwee Meadows, 68
Haiwee Reservoir, 168
Halloran Springs overpass,
 209–10
Halloran Summit, 208–10
Halloran Wash, 210
Hanaupah fan, 112–15
Hanaupah scarp, 113–14
Hancock Park, 31
Harmony borax plant, 98
Harper Lake fault, 41
Harry Wade Historical
 Monument, 97
Harvard Road, 86–87
Hasley Road, 245–46
Helendale fault, 41, 82
Hesperia–Phelan exit, 79
Hidden Spring picnic area, 134
High Divide, 59
Highland Springs turnoff, 231
Hilton Creek glacier, 198
Hilton Creek moraine, 198
Hodge Road, 82–83
Hollow Hills, 92, 208
Holser fault, 245
Homestead, 159
Hot Creek, 201
Hot Mineral Spa, 47
Hungry Valley, 249
Hungry Valley Formation (beds),
 253–54
Hungry Valley Road, 253
Huntington Hotel, 33
Huntington Library, 33

I

Ibex Mountains, 97–98
Ibex Pass, 97–98
Imperial (town), 46
Imperial fault, 46

Imperial Formation, 44–45
Imperial Irrigation District, 47
Imperial Valley
 gypsum mining in, 28
 hot brines under, 47
 lake deposits in, 45
 unconsolidated fill under, 44
Independence, 179–80
Independence Creek, 186
Indian Avenue, 236
Indian Wells, 159
Indian Wells Valley, 128
 Owens Lake emptying into, 68
 routes converging in, 119
 view of, 158–59
Indio Hills, 46
Inface bluffs, 78
Inglewood fault, 32–33, 55
Interstate Freeway 215, 228
Interstate Highway 5
 Bucket of Worms interchange
  of Highway 14 and,
  137, 242
 fault scarp between lanes
  of, 257
 segment T from San Fernando
  Pass to San Joaquin Valley
  along, 242–58
 Transverse Ranges divided
  by, 20
Interstate Highway 8,
 experimental power plants
  off of, 51
Interstate Highway 10, 230
Interstate Highway 14, 119, 136
 approach to Bishop on, 149
 Bucket of Worms interchange
  of Highway 5 and, 137, 242
 ridges passed by, 143
 syncline on, 242
Interstate Highway 15
 bedrock knobs of granitic rock
  along, 83
 separation of U.S. Highway
  395 and, 79–80
Inyo craters, 62, 206

Index

Subject Index

A

Abrasion (wind), 116, 235–37
Aftershocks, 238
Age of the Earth, 259
Agglomerates, volcanic, 143
Alkali salts, 249
Alluvial apron, 119, 167
Alluvial cones, 106
Alluvial divide, 42
Alluvial fans
    apron of, in Los Angeles Basin
        to San Gabriel and San
        Bernardino mountains, 79
    at base of Black Mountains,
        105, 115
    in the Artists Drive block, 116
    in the Basin Ranges province,
        65, 68
    breakup of earlier Mojave
        River flow by, 42
    in Death Valley, 104, 107,
        109–11
    east of Avawatz Mountains, 95
    fault scarplets cutting, 67, 107,
        114–15
    in the Great Valley province, 56
    human settlements on, 239
    in Owens Valley, 63, 179
    from Sierra Nevada, 179
    in Taboose-Big Pine volcanic
        field, 182
    transportation of, 161

Alluvial fill
    in Basin Ranges, 67
    in Death Valley, 106
    in Owens Valley, 188
    in western Mojave Desert, 38
Alluvial gravels, 78, 98, 107, 161
Alluvial plains, 56, 86, 234, 236
Alluvial slopes, 129
Alluvian apron, 119
Alluvium
    basins of, 29–30
    defined, 7, 20
    in the Mojave Desert, 38, 41
    as source of sand, 95
    weathered, 136
Ancient humans, 15
Angular unconformity, 246
Anorthosite, 134–35, 142
Antecedent streams, 36–37
Anticlinal folding, 258
Anticlines
    beds of sand and gravel folded
        into, 258
    in Death Valley, 102
    defined, 4
    in the Great Valley, 56–58
    in the Los Angeles Basin,
        33–34
    oil migration into, 22
    plunging, 33–34, 109
    recognizing, 7
    young, 83
Antimony mined in Mojave
        Desert, 42
Archeological digs, 86
Archeology, 227

Artifacts, 227
Arvin-Tehachapi earthquake
    (1952), 57
Ash, volcanic, 43, 194
Axis defined, 32

*Austin, Mary "Land of Little Rain"
of Independence*

B

Badlands
    in Bryce Canyon (Utah), 229
    in inland basins, 28
    in Red Rock Canyon, 156
    in San Timoteo Formation,
        228–29
    in Tecopa lake beds, 98
Banks, stream, 198
Barchan dunes, 50
Barite, 63
Barranca, 253
Basalt, 116
Basaltic lavas, 3, 182, 201
Base line, 224
Basement relief, 32
Basin fill, 31
Bastnaesite, 212–13
Batholith
    defined, 26
    in Peninsular Ranges province,
        27–28, 228
    Sierra Nevada, 61–62
Beach gravels, 90
Beach ridge, 88–89
Beaches, causes of sand loss
        on, 15